The Nature of Difference

Science, Society and Human Biology

Society for the Study of Human Biology Series

10 Biological Aspects of Demography
Edited by W. Brass

11 Human Evolution
Edited by M. H. Day

12 Genetic Variation in Britain
Edited by D. F. Roberts and E. Sunderland

13 Human Variation and Natural Selection
Edited by D. Roberts (Penrose Memorial Volume reprint)

14 Chromosome Variation in Human Evolution
Edited by A. J. Boyce

15 Biology of Human Foetal Growth
Edited by D. F. Roberts

16 Human Ecology in the Tropics
Edited by J. P. Garlick and R. W. J. Keay

17 Physiological Variation and its Genetic Base
Edited by J. S. Weiner

18 Human Behaviour and Adaption
Edited by N. J. Blurton Jones and V. Reynolds

19 Demographic Patterns in Developed Societies
Edited by R. W. Horns

20 Disease and Urbanisation
Edited by E. J. Clegg and J. P. Garlick

21 Aspects of Human Evolution
Edited by C. B. Stringer

22 Energy and Effort
Edited by G. A. Harrison

23 Migration and Mobility
Edited by A. J. Boyce

24 Sexual Dimorphism
Edited by F. Newcome et al.

25 The Biology of Human Ageing

26 Capacity for Work in the Tropics
Edited by K. J. Collins and D. F. Roberts

27 Genetic Variation and its Maintenance
Edited by D. F. Roberts and G. F. De Stefano

28 Human Mating Patterns
Edited by C. G. N. Mascie-Taylor and A. J. Boyce

29 The Physiology of Human Growth
Edited by J. M. Tanner and M. A. Preece

30 Diet and Disease
Edited by G. A. Harrison and J. Waterlow

31 Fertility and Resources
Edited by J. Landers and V. Reynolds

32 Urban Ecology and Health in the Third World
Edited by L. M. Schell, M. T. Smith and A. Billsborough

33 Isolation, Migration and Health
Edited by D. F. Roberts, N. Fujiki and K. Torizuka

34 Physical Activity and Health
Edited by N. G. Norgan

35 Seasonality and Human Ecology
Edited by S. J. Ulijaszek and S. S. Strickland

36 Body Composition Techniques in Health and Disease
Edited by P. S. W. Davies and T. J. Cole

37 Long-term Consequences of Early Environment
Edited by C. J. K. Henry and S. J. Ulijaszek

38 Molecular Biology and Human Diversity
Edited by A. J. Boyce and C. G. N. Mascie-Taylor

39 Human Biology and Social Inequality
Edited by S. S. Strickland and P. S. Shetty

40 Urbanism, Health and Human Biology in Industrialized Countries
Edited by L. M. Schell and S. J. Ulijaszek

41 Health and Ethnicity
Edited by H. Macbeth and P. Shetty

42 Human Biology and History
Edited by M. T. Smith

43 The Changing Face of Disease: Implications for Society
Edited by C. G. N. Mascie-Taylor, J. Peters and S. T. McGarvey

44 Childhood Obesity: Contemporary Issues
Edited by Noel Cameron, Nicholas G. Norgan, and George T. H. Ellison

45 The Nature of Difference: Science, Society and Human Biology
Edited by George T. H. Ellison and Alan H. Goodman

Numbers 1–9 were published by Pergamon Press, Headington Hill Hall, Headington, Oxford OX3 0BY. Numbers 10–24 were published by Taylor & Francis Ltd, 10–14 Macklin Street, London WC2B 5NF. Numbers 25–40 were published by Cambridge University Press, The Pitt Building, Trumpington Street Cambridge CB2 1RP. Further details and prices of back-list numbers are available from the Secretary of the Society for the Study of Human Biology.

The Nature of Difference

Science, Society and Human Biology

Edited by

George T. H. Ellison • Alan H. Goodman

The Wenner-Gren Foundation

For Anthropological Research, Inc.

CRC Press
Taylor & Francis Group
Boca Raton London New York

CRC Press is an imprint of the
Taylor & Francis Group, an **informa** business
A TAYLOR & FRANCIS BOOK

wellcome trust

Cover image created by Emma Stone.

Published in 2006 by
CRC Press
Taylor & Francis Group
6000 Broken Sound Parkway NW, Suite 300
Boca Raton, FL 33487-2742

First issued in hardback 2018

ISBN 13: 978-1-138-40386-4 (hbk)
ISBN 13: 978-0-8493-2720-9 (pbk)

Library of Congress Cataloging-in-Publication Data

Catalog record is available from the Library of Congress

informa
Taylor & Francis Group
is the Academic Division of Informa plc.

Visit the Taylor & Francis Web site at
http://www.taylorandfrancis.com

and the CRC Press Web site at
http://www.crcpress.com

PREFACE

Diversity, Difference, Deviance: Ethics and Human Biology

The chapters in this volume were drawn from papers presented at the 45th Annual Symposium of the Society for the Study of Human Biology, sponsored by the Wenner-Gren Foundation and the Wellcome Trust's research program in Biomedical Ethics. The symposium took as its theme 'Diversity, Difference, Deviance: Ethics and Human Biology' and aimed to explore:

(i) the ethical issues that accompany the scientific production of data on human biological variation; and
(ii) how human biologists might tackle the social, political, and ethical consequences of their research

The symposium solicited papers from a range of different academic disciplines to generate a rich tapestry of contrasting, yet essentially complementary, perspectives on human nature, the nature of difference, and related scientific practice.*

The chapters included in this volume offer a detailed analysis of selected presentations. These were chosen to represent the range of issues addressed, and to facilitate a synthesis of the multi-disciplinary discourse and inter-disciplinary dialogue the symposium was able to provoke. At the same time, this selection of chapters, and the way they have been arranged in this volume, sought to address a number of specific suggestions and concerns raised by symposium participants during an open forum at the end of the final day.

---

* Abstracts and extended summaries of papers and posters presented at the three-day symposium have been published in a special issue of the BioSocial Society's in-house journal, *Society, Biology and Human Affairs*. 2002; 67(2).

Participants at the 45th Annual Symposium of the Society for the Study of Human Biology on 'Diversity, Difference, Deviance: Ethics and Human Biology'

Many participants told us that they found the symposium's diverse range of presentations, and the opportunities these provided for synthesizing contrasting disciplinary perspectives, both refreshing and important. They were refreshing because the multi-disciplinary discourse and inter-disciplinary dialogue offered by contributors to the symposium provided unexpected insights into biological explanations for human nature and difference. Important because the symposium demonstrated that different disciplines *could* talk to one another and that, despite their more obvious differences, there was extensive common ground. The symposium provided a tangible sense that bringing together different perspectives on human biology and the nature of difference enabled progress toward characterizing many of the conceptual and ethical problems that human biologists face.

Nonetheless, some participants expressed concern about the possible marginalization of biologists themselves. They felt that an interdisciplinary lens which explored scientific practices risked being misinterpreted as an 'anti-science' agenda which criticized and de-legitimized 'scientific' topics and analytical approaches on essentially political grounds. This was a valid concern, not least because some of the contributors from the biological sciences were wary of being seen as token representatives of their craft.

We took this potential risk seriously, not least because many of the participants from non-biological disciplines responded by describing how they too had felt marginalized during parts of the symposium. However, they suggested that everyone probably felt as if they were sometimes "standing outside" of the symposium, if only because the range of perspectives and the inter-disciplinary dialogue was such unfamiliar territory. Indeed, there was consensus that it was important to acknowledge and address the potential to alienate some academic constituencies. The participants encouraged us to ensure that all relevant disciplines were adequately represented within the proceedings, and that the volume was accessible to students and a wider audience beyond the confines of academia.

In constructing this volume we sought to achieve these ancillary objectives, by selecting: contributors from a wide range of different disciplinary backgrounds – including the humanities, the social and natural sciences, and disciplines in-between (such as bioethics); and contributions on a wide range of relevant topics and perspectives — including the material, philosophical and symbolic nature of 'difference,' and the popular and scientific practices that inform our understanding thereof. In the process the following chapters demonstrate the extraordinary scope of contemporary research and scholarship on the nature and meaning of human biological differences, and the important contribution that different disciplinary perspectives can bring to this work.

Because diverse analytical styles and foci tend to make such compilations vulnerable to inaccessible or incomprehensible eclecticism — accentuating the very differences this volume aims to transcend — each of the contributors was tasked to reach out beyond their own community of practice to engage those with very different perspectives and very different analytical approaches. For similar reasons, the chapters have been carefully arranged into three sections: "looking for 'nature' in human nature;" "geneticization and the nature of difference;" and "scientific practice and the pursuit of difference" – each with four chapters and with 'bridging' chapters at the beginning and end of each section. The volume's three sections aim to combine different disciplinary perspectives in a way that challenges popular and conservative paradigms – paradigms that undermine and constrain our ability to understand the nature and causes of difference.

The Nature of Difference: Science, Society and Human Biology

The volume has thereby evolved into a collection of perspectives on the "nature of difference" – what difference *is*; what difference *means*; and what *causes* difference. As such the contributions selected for inclusion herein demonstrate how the different perspectives from unrelated disciplines (which are, after all, subcultures of the wider academic community) influence the answers to each of these questions. To some, this variation and lack of consensus might seem to question the validity of what each of us classify as 'biological,' and might even undermine the importance of biology as a determinant of difference. This more critical analysis of different disciplinary perspectives might also adopt a less optimistic view of the volume's three sections which they might label: "the epistemological importance and theoretical power of biology-as-nature;" "the social recognition, reification and 'naturalization' of difference;" and "the corruption of scientific enquiry by social perspectives of difference." Yet this is essentially a cynical approach that views the lack of consensus as problematic rather than illuminating. Instead, we aimed to demonstrate the potential benefits of combining different disciplinary perspectives which challenge the popular and conservative paradigms that undermine our ability to understand and address the nature and causes of difference.

The first section examines the role that 'nature' plays in contemporary explanations for human nature. It opens with Cartmill's thoughtful essay exploring the arguments for and against a biological basis for morality and concluding that language provides the crucial link between our biological and social selves, and between our material, philosophical and symbolic nature. This is followed by Dingwall et al.'s analysis of past attempts to situate human behavior within a biological framework. They use criminality as an example to revisit the sociological critiques of biological determinism and the limits these critiques face without a rapprochement between the social and natural sciences. This is evident in Marks' subsequent chapter, which explores the symbolic and scientific value of comparisons between humans and non-human primates. He concludes that we need to recognize the capacity of such aesthetic comparisons to constrain and subvert scientific analyses. Finally, the section concludes with Durrheim and Dixon's empirical study of popular explanations for biological and social difference in South Africa. They demonstrate how scientific methods and arguments are co-opted within lay ontologies which reify and naturalize difference and provide tautological explanations of difference that reflect and reinforce illegitimate scientific analyses.

The second section of the book includes four chapters that consider the important role that genetics has come to play in naturalistic explanations of human difference – a role that imposes a natural cause and individual responsibility for difference through a process that has become known as 'geneticization.' In this regard, Wexler's chapter follows on from Durrheim and Dixon by demonstrating how popular representations of 'witchcraft' have been subject to crude re-interpretation and put to use for eugenicist ends by what appears to have been a deliberately biased genealogy of Huntington's chorea. This was a genealogy that was socially and scientifically satisfying because it successfully mapped the stigma of disease against the evils of witchcraft to reproduce the imaginary material and metaphysical symmetry of deviance. The role of genetics in sexuality forms the focus for the next chapter, in which Kitzinger compares media and scientific representations of an apparent association between male homosexuality and specific genetic traits. The media questioned these 'findings' on various grounds, particularly the basis on which social phenomena were: deemed to require a 'cause;' selected as relevant for scientific analysis; and thereby reified as materially different. Shakespeare takes up these points in his analysis of the role of social and biological factors in defining and producing 'disability.' He suggests that research into the genetic basis of disability does not explain why such research is necessary or relevant, while it ignores the social production and interpretation of difference as *dis*-abled. Finally, Ashcroft demonstrates how race and ethnicity have come to play such an important role in the application of research into genetic determinants of drug response (so-called 'pharmacogenetics') despite serious social and scientific misgivings. This is a practice that seems to reflect the way in which social and scientific perspectives work together to satisfy preconceived interpretations of difference and thereby conserve popular biological explanations of human difference.

The third and final section of the book takes a closer look at how research into human biological difference is practiced. Outram and Ellison's chapter follows on from Ashcroft's analyses by exploring the views and experiences of geneticists and genetics journal editors concerning the use of race and ethnicity in genetic research. They suggest that geneticists implicitly recognize the questionable reliability, validity and sensitivity of racial and ethnic categories as markers of genetic difference, but separate their pragmatic use of these (as rough and ready scientific tools) from their social meanings and consequences. This is an issue Santos considers when exploring the ethical challenges facing genetic and related biological research on isolated human populations. By examining the regulatory constraints that govern human biological research on such populations, using Brazilian Amerindians as a case study, he is able demonstrate how these continue to draw on historical caricatures of difference, and adopt a protectionist rather than an emancipatory approach which stifles research into human diversity. Not that all such regulation need operate in this fashion – as evident in the next chapter by Turner et al. They review a range of (un)ethical research practices in skeletal biology and forensic anthropology, past and present, and conclude that substantial scientific benefits have accrued from recent legislation (such as NAGPRA in the USA) and related developments which help scientists use biological material as evidence of social and cultural processes. The book concludes with Goodman's call to arms, which argues forcibly for the reintegration of physical/biological and social/cultural anthropology, drawing on the examples of race-as-biology and stature to demonstrate why it is impossible to understand human biology outside its socio-cultural context.

Beyond their collective focus on the role of science, society and human biology in defining the nature of difference, the chapters in this volume raise three additional issues that have wider ramifications for scientific research into human biology and beyond. The first of these involves the tautological tendencies of all scientific enquiries. The second concerns the exceptional power of biological explanations for 'difference.' The third considers what science and society might do to escape the clutches of biological determinism. All are issues that are addressed in greater depth in the introductory chapter.

George T. H. Ellison and Alan H. Goodman

ABOUT THE COVER

Representing the material and symbolic in the Nature of Difference

Creating a cover to reflect the complex and eclectic nature of this volume was always going to be a challenge – not least because it draws on such a fundamental concept (human nature) and on the ways in which the biological and social sciences offer such different yet complementary perspectives on: what it is to be human; what it is to be different; and what lies at the heart of human nature. The cover graphic ended up being done in the garden (which was really overgrown at the time) – a context that presents the figures as if they were in a village or their own habitat. Pink was used to symbolize the color of flesh, and to link the figures and their context with an organic hue. However, the only things that are 'properly' pink are the figures, while the use of pink elsewhere serves to emphasise a recurrent theme of the book – aberration and deviance as a 'normal', or at least an 'ordinary,' feature of human biological diversity.

As with my other work, I sought to create not only a larger composite graphic, but one filled with lots of smaller images, so that the closer you look the more you see and the greater the evident complexity. In this way the cover graphic is meant to show you something new, something you didn't see before, every time you look at it. The choice of dolls is deliberate – as powerful yet strange caricatures of human beings they are often not very likable and, in the ways they are presented here, they are intended to be perverse and ostensibly dead. When we die and are buried we are still intact as organic doll-like figures, but when we look again at *these* ostensibly dead figures (who are neither dead nor alive) they appear soul-less and eerie, more like artifacts or exhibits in a museum. As such they mimic the objectified remains which are the core material of many of the human biologists and social scientists who have contributed to this volume. From what they have to tell us it is clear that we still have so much more to learn, particularly about who we are, what we are, what makes us different, what makes us the same, and how we are all unified in death.

Russian dolls were chosen because they have smaller versions of themselves *within* themselves, reflecting how each of us developed as unique individuals, and how humans evolved as a species. The animals came into the picture, where they are half-animal and half-(Russian) doll, to signify our ancestry as well as deformity and deviance – suggesting that we are all, in some way or another, 'mutants.' Some of the figures are broken up (like the cow coming out of the base of one doll, and the dog who looks as if he is growing the wrong way out of another doll, almost like a 'bad egg'), and again this is meant to signify the inner animal or beast within emerging as part and parcel of human nature. Likewise, the half-sheep-half-Russian doll (which almost looks like a Cabbage Patch doll), sets out to illustrate how it is we come about – how we're born, how we come of age and of what are we made. This is perhaps most evident in the figure (bottom right) who has as his stomach: the sea; Russian dolls; two tiny cows; a sheep; and a smaller version of himself – all deep inside him. Again, what this aims to signify is that inside all of us are leftovers from our ancestors and predecessors – a heady brew that many feel determines our nature.

I was particularly drawn to using the sea, in part to reflect the salty fluid in which all life exists, in part to touch on the chapter exploring social explanations for racial segregation on beaches during apartheid. But using the sea also aims to signify the 'factual' – the sea is unchanging and is, in many respects, a constant in a world of change. Meanwhile the chapter on segregation is also highlighted by the separation of the two sheep and two pigs – in this way the cover graphic seeks to demonstrate, and reiterate through repetition, that different 'types' can be distinct or can be forced to be distinct by those around them. And as the figures in the cover graphic came together they all seemed to be looking at each other (particularly the five dolls in a row top left, who are all looking at the half-doll/half-sheep), yet they're not communicating. Indeed, they are in one way or another related but disengaged from one another – an arrangement that aims to symbolize the common strands of research into the nature of difference from the biological and social sciences, yet the difficulties these face when trying to communicate with each other.

Meanwhile, I particularly like the shoe because it's bigger than everything else, even the car (which should, of course, be much larger). This sets up a contradiction, where things that are supposed to be big are small and vice versa, to remind us that nothing is certain – particularly how individual humans have come about, developed and grown in unique and individual ways. This reflects discussions within the book between biologists and social scientists, the former wanting to understand the world as hard and fast, the latter wanting to explore the world's contradictions and paradoxes. Neither seem entirely sure that either approach provides an adequate view of the whole picture, and instead seem to accept the importance of adopting more contingent and flexible perspectives to fully understand the complexity of human nature – just as we do when exploring the cover graphic.

In summary, then, it is a design that aims to reflect complexity and the rather flat and static material and factual basis through which human difference is often explored – yet alongside are the more spiritual aspects of human nature, and this is evident in the way the picture is both disturbing and disquieting. Indeed, there is an innocence on the surface (emphasised through the use of dolls and toys, such as the cutesy little mouse) which is intended to be subversive with a sinister quality underneath: the use of children's toys in any setting other than a playroom always feels sinister, and the fact that in the cover graphic they are in the garden (and thereby out in the open) creates a tension – a sense that they or the child who owns them is in danger – a theme which draws on the risks we all face when abandoning or exploring too deeply our inner nature.

Emma Stone – cover artist

ACKNOWLEDGMENTS

This volume would not have been possible without the support of Nick Norgan who was president of the Society for the Study of Human Biology when the 45th Annual Symposium took place. Nick encouraged us to go beyond the traditional boundaries of the society's symposia clientele, to draw on perspectives from the social sciences and humanities to better understand what it is that human biologists study and what it is that human biologists do. This volume aims to be a tribute to his enthusiasm for this unconventional approach and we hope it does him proud.

In addition to the support provided by the Society for the Study of Human Biology, we are also grateful to the substantive funds provided by the Wenner-Gren Foundation (whose support enabled us to invite academics from outside Europe and the US) and by the Wellcome Trust's biomedical ethics research program (whose financial assistance enabled us to invite more of our colleagues from the US and UK). Taylor & Francis kindly sponsored the reception on the opening evening of the symposium, and the Institute of Education at the University of London provided an excellent venue.

Most important of all we are grateful to the contributors to the symposium including those whose presentations are not included in this volume (Michael Blakey, David Braunholtz, David Gresham and Luba Kalaydjieva, Anne Kerr, Sheila Maclean, Kenan Malik, Hilary Rose, and Steven Rose) and those who acted as additional discussants to each of the symposium sessions (Laura Bishop, Bob Connolly, SH Cedar, Jeanelle de Gruchy, Sarah Elton and Catherine Panter-Brick). All played an important part in the content of this volume and we trust that the presentations we were able to include do justice to the issues they raised and the contributions they made.

Finally, we would like to thank a number of friends and colleagues who helped us pull the volume together, in particular: Gill Mein, Chaia Heller,

Thea de Wet, Jane Wills, Mark Gilthorpe, Yvonne Erasmus and Mary Halter. The volume would not have been completed without the support of the Taylor & Francis staff, especially Kevin Craig, Alex Levin, Barbara Norwitz, and Amber Stein and, in particular, we owe a huge debt of thanks to Pat Roberson for steering this through to the press.

ABOUT THE EDITORS

George T. H. Ellison is currently professor of health sciences at St. George's, University of London, where his interests focus on the use of race and ethnicity in biomedical research, social determinants of health inequalities, lifecourse epidemiology, and evidence-based decision-making. Following a B.Sc. in zoology at the University of Aberdeen and a Ph.D. in comparative physiology at the University of Pretoria, George taught applied biology at the University of Greenwich while studying for his master's degree in health promotion at the London School of Hygiene and Tropical Medicine. After two years back in South Africa, working first as senior researcher with Thea de Wet on the Birth to Ten study and then as research manager for the Institute of Urban Primary Health Care at Alexandra Health Centre and University Clinic, he returned to the United Kingdom as a lecturer in the Department of Biological Anthropology at the University of Cambridge. He subsequently took up the post of assistant director in Ann Oakley's Social Science Research Unit at the Institute of Education before joining London South Bank University as inaugural director of the Institute of Primary Care and Public Health, from which he moved to St. George's in 2004.

During this period George's research evolved from comparative physiology to nutrition, and from maternal and child health to a focus on health services research. More recently he has been drawn to interdisciplinary studies in social epidemiology and medical anthropology by the unique perspectives such "boundary disciplines" bring to understanding the interaction between social and biological processes. Previous edited books include: G.T.H. Ellison, M. Parker and C. Campbell (Eds.) *Learning from HIV and AIDS* (2003, Cambridge University Press); and Cameron N., N.G. Norgan and G.T.H. Ellison (Eds.) *Childhood Obesity: Contemporary Issues* (2005, CRC Press). His current research includes a Wellcome Trust-funded interdisciplinary project with colleagues from Imperial College,

Nottingham University, and the University of Oxford, exploring the role of race and ethnicity in genetic research and the development of pharmacogenetic technologies. He lives in South London with his partner, the social scientist Gill Mein, and their son, the aspiring veterinarian Ralph.

Alan H. Goodman is the current president of the American Anthropological Association, the world's largest professional organization of anthropologists, and professor of biological anthropology at Hampshire College, Amherst, Massachusetts, where he is also the former dean of natural sciences. In addition to receiving a B.S. (Magna Cum Laude) in zoology and a Ph.D. in anthropology from the University of Massachusetts, Amherst, he has been a research fellow at the Laboratory for Stress Research, Karolinska Institute, Stockholm and a postdoctoral student in international nutrition at the University of Connecticut.

Goodman's research and teaching focus on the intersections among political–economic processes, ecology, ideology, and human biology. Particular research foci include the biological consequences of racializing in fields such as public health, forensics, and genetics and how commodities such as junk foods and soft drinks have consequences for nutrition and health status. He has published over 100 papers on issues such as the persistence of the idea of race in science and the health consequences of unequal access to resources in the past, particularly as indicated by developmental defects and chemical changes in tooth enamel. Previous edited books include A.H. Goodman and T. Leatherman (eds.), *Building a New Biocultural Synthesis: Political Economic Perspectives in Biological Anthropology* (1998, University of Michigan Press); A.H. Goodman, et al. (eds.) *Nutritional Anthropology: Biocultural Perspectives on Food and Nutrition* (2000, McGraw Hill); and A.H. Goodman, et al. (eds.) *Genetic Nature/Culture: Anthropology and Science Beyond the Two-Culture Divide* (2003; University of California Press). He is an associate director of the New York African Burial Ground project and the recipient of numerous grants from the Wenner-Gren Foundation, the National Institutes of Health, and the National Science Foundation. In 2000, Goodman received The World of Difference Teaching Award from the Anti-Defamation League.

CONTRIBUTORS

George J. Armelagos
Department of Anthropology
Emory University
Atlanta, Georgia, USA

Richard Ashcroft
Medical Ethics Unit,
Department of Primary Care and Social Medicine
Imperial College
London, UK

Matt Cartmill
Department of Biological Anthropology and Anatomy
Duke University Medical Center
Durham, North Carolina, USA

Robert Dingwall
Institute for the Study of Genetics, Biorisks and Society
University of Nottingham
Nottingham, UK

John Dixon
Department of Psychology
University of Kwazulu-Natal
Durban, South Africa

Kevin Durrheim
Department of Psychology
University of Kwazulu-Natal
Durban, South Africa

George T. H. Ellison
St. George's University of London
London, UK

Alan Goodman
Department of Natural Science
Hampshire College
Amherst, Massachusetts, USA

Samantha Hillyard
Institute for the Study of Genetics, Biorisks and Society
University of Nottingham
Nottingham, UK

Jenny Kitzinger
Cardiff School of Journalism, Media and Cultural Studies
Cardiff University
Cardiff, Wales, UK

Jonathan Marks
Department of Sociology and Anthropology
University of North Carolina
Charlotte, North Carolina, USA

Brigitte Nerlich
Institute for the Study of Genetics, Biorisks and Society
University of Nottingham
Nottingham, UK

Simon M. Outram
St. George's, University of London
London, UK

Ricardo Ventura Santos
Escola Nacional de Saude Publica
Rio De Janeiro, Brazil

Tom Shakespeare
International Centre for Life
University of Newcastle
Newcastle, UK

Diana S. Toebbe
Department of Anthropology
Emory University
Atlanta, Georgia, USA

Bethany L. Turner
Department of Anthropology
Emory University
Atlanta, Georgia, USA

Alice Wexler
Center for the Study of Women
University of California Los Angeles
Los Angeles, California, USA

CONTENTS

Part 3 Scientific Practice and the Pursuit of Difference

INTRODUCTION

Human Biology and the Embodiment of Difference

George T. H. Ellison
St. George's, University of London

The Nature of Difference: Science, Society, and Human Biology

From its very inception, the field of human biology has been the quintessential science of variation and difference. Francis Galton, Charles Darwin's cousin, set the stage for the statistical analysis of normality and deviance (Gillham, 2001). Aleš Hrdlika, the founder of the *American Journal of Physical Anthropology*, famously pronounced in that journal that the goal of physical anthropology is the study of "the normal white male." But difference is more than statistics and more than just biology. What we see as difference, and similarity too, is also the result of social conditions and social conditioning. In the United States, for example, variation in skin color seems impossible to ignore. But we could hardly imagine being as aware of variation in, say, the shape of one another's ears. Difference is social, and how it is studied and theorized has a number of consequences for scientific studies and for the peoples we are studying. Despite the centrality of difference throughout the history of research into human biology, from craniometric variation to genomic diversity, 'difference' itself has rarely been the focus of our attention. What, we might ask, are some of the ethical, social, political, and scientific issues in how we conceptualize, analyze, and theorize difference itself?

To this end the four chapters in the first section of this volume provide a compelling insight into the scientific and social arguments that situate differences in human nature within biological processes. The chapters in the second section go on to explore the role that genetic findings and genetic mechanisms have come to play in these arguments. The chapters in the third section conclude the volume by examining the scientific structures and practices that are responsible for and sustain this approach.

Section 1: Looking for 'Nature' in Human Nature

In the very first chapter in this volume Matt Cartmill explores the contribution biology might make to universal moral imperatives by revisiting the last great attempt to ground morality in biology: E. O. Wilson's (1975) proposal that morality was based on natural selection. Cartmill tackles each of Wilson's propositions in turn. First, like many contemporary anthropologists, he questions the evidence available for biological (i.e., genetic and neurological) mechanisms that control complex human behaviors, and suspects that conclusive evidence will never (and, in fact, can never) be found. Next, he examines a more substantive concern: that even if biology were to explain the way we behave it would not explain the way we felt we ought to behave. Biological mechanisms, even those developed through natural selection as adaptive traits to improve reproductive fitness, are not necessarily "good." Indeed, viewing the biological consequences of natural selection as inherently good is simply a sophisticated form of the naturalistic fallacy. In particular, Cartmill points out that because "selective advantage can accrue to wicked impulses as well as virtuous ones . . . then it cannot furnish criteria [or be the basis] for distinguishing between the two."

Adaptive considerations cannot be relied upon to explain the basis of morality because "there is no law of nature that prevents adaptively useful impulses from conflicting with each other." Thus, at least from the naturalistic perspective of Wilson and the evolutionary psychologists who inherited the mantle of sociobiology, Cartmill does not believe that adaptationist arguments provide a compelling case for a biological basis of morality. Instead, Cartmill proposes that morality might be an epiphenomenon of language, a universal yet uniquely human trait. Language allows two other processes to occur—logical argument and self-awareness—which in turn allow humans to make judgments and consider the consequences of these judgments on others. Yet if language is the basis of morality, morality has a foundation in human nature that is, from a materialist's perspective, biological. Language gives us a place outside ourselves to stand and make judgments about the self, to shift from first to third person, from subjective to objective, and back again. It gives us the ability to look at ourselves from the inside and from the outside, and although we may never do so with total objectivity, language and awareness at least provide a mechanism that enables us to recognize this.

The second chapter in this section, by Robert Dingwall, Brigitte Nerlich, and Samantha Hillyard, also focuses on social science critiques of attempts to construct biological explanations for complex human behaviors. As an example they draw on Lombroso's (1911) biological explanations for criminal behavior which drew on the Darwinian notion of 'atavism': a behavioral reversion to subhuman savagery, identifiable from physical features similar to those of early hominids or lesser primates. Although these ideas have been discarded, the notion of atavism remains central to contemporary theses linking biology to behavior, while the focus on gross morphological or psychological features has shifted to the genetic level. Nonetheless, a focus on associations between genetic and behavioral traits suffers from the same methodological weaknesses as foci on morphological and psychological associations. These weaknesses relate, primarily, to a tendency for biologists to oversimplify their characterization and measurement of "*the* environment," a term that often excludes the notion of *social* environment*s*. Moreover, most biologists fail to grasp that their attachment to 'realism' is not immune from social processes that confirm what is 'real' and provide collective reassurance of such 'facts.' Thus to simplify or reduce human behavior to the instinctive interactions of lower animals is to ignore the symbolic value and power of language.

For social scientists, the whole notion of a biologically determined act (such as a genetic predisposition for criminality) depends upon a collective understanding of what that act is. Although this line of postmodernist reasoning can lead some to reject the reality of the body (and, likewise, the gene) as essential to the experience of self and the expression of behavior, Dingwall et al. point out that experience is embodied simply because this is what makes experience possible, and embodiment then provides phenomena that are open to social interpretation. If there is to be a rapprochement between the biological and social world views, Dingwall et al. propose three issues that must be addressed. Both sides should actively engage with each others' arguments and strive to understand and respect each other's perspectives. At the same time, biologists must engage with the inherent diversity of what they term *the* environment, going beyond the physical conditions familiar to ecologists and epidemiologists, to consider the important cultural and symbolic roles that environment*s* (both physical and social) play in the expression and experience of biological difference. Finally, social scientists need to recognize the unique insights that biological research provides by exposing the embodiment of our social worlds.

Following on from Dingwall et al.'s examination of atavism, Jonathan Marks explores the dual role that nonhuman primates play in anthropological analyses: first, as natural objects in the comparative analyses undertaken by physical anthropologists, and second, as cultural symbols in the sorts of rhetorical comparisons examined by social anthropologists. These two very different views of nonhuman primates mimic the parallel yet contradictory views of evolutionary theory. Evolutionary theory operates as a robust, and essentially neutral, challenge to creationist dogma. On the other hand it is a tool that has been co-opted and corrupted by racists, eugenicists, and other soci(obiologic)al engineers. Describing people as 'monkeys' has been a popular jibe for hundreds of years, and both popular and scientific comparisons between humans and nonhuman primates were made long before Darwin published his theory of evolution. It is therefore unclear what function these comparisons might have served in the absence of a unifying biological consensus (i.e., before the advent of evolutionary theory).

Marks suggests that, even for biologists, these comparisons were and are symbolic rather than scientific. For example, early visual/diagrammatic representations of different human races alongside nonhuman primates were intended to illustrate esthetic, rather than ancestral anatomical/biological, relationships. Marks therefore argues that scientists subsequently and "readily appropriated the familiar imagery of pre-Darwinian racial thought" and used these comparisons between nonhuman primates and 'lesser' races as a "tool for dehumanizing" these 'others.' Thus, far from being a useful scientific tool, the comparison of humans and nonhuman primates serves primarily as a "rhetorical tool for the symbolic dehumanization of unfavored people." This is surprising given that, on the face of things, scientific comparisons between humans and apes should be fruitful, inasmuch as humans and nonhuman primates share what Marks calls "an intimate biological history." Yet comparisons of similar processes in humans and apes (such as rape and infanticide) tend to invoke biological homology even when the processes manifest very differently and the comparison between them better reflects phonological homonyms: a tendency to imbue nonhuman primates with human qualities in essentially unhelpful, anthropocentric analyses. Nonetheless, Marks concludes that we cannot study human biological diversity without comparing humans to nonhuman primates. But he cautions that anthropologists must be sensitive to the cultural messages such comparisons convey because nonhuman primates are cultural constructions used to diminish the humanity of the people to whom they are compared.

These sorts of comparisons form an important part of the quasi-scientific explanations for racial segregation provided by the white South Africans interviewed by Kevin Durrheim and John Dixon for the concluding chapter in this section. Durrheim and Dixon draw on Moscovici's (1948) work to explore how commonsense explanations of human difference use scientific findings and scientific arguments to sustain a racialized view of society that transcends stereotyping or prejudicial tendencies. They argue that the contemporary popularization of science in everyday discourse has encouraged and enabled individuals and publics to theorize and construct their own quasi-scientific rationales for the world they see around them (and the attitudes they experience). However, by focusing on rhetorical interactions, rather than explanations per se, Durrheim and Dixon demonstrate that such rationales are flexible and that individuals draw on a variety of explanations to justify, modify, and refine their views or understanding of human difference. Thus, although these explanations take the form of what Durrheim and Dixon call "lay ontologizing" (i.e., arguments based on quasi-scientific principles and methods) and draw on the biological bases for human difference, they also draw on cultural and transcendental perspectives, often during the same conversation. Nonetheless, these explanations tend to be based on little more than a plausible yet tautological argument that things are "the way they are" because that's "the way they are" . . . an argument that relies on a variety of different views of "nature" (part biological, part cultural, part transcendental).

Seeing things "the way they are" as "natural," or as evidence of "human nature," creates an innate and essentialized basis for prejudice and stereotyping that does not require (but occasionally draws on, or develops into) biological arguments. Indeed, despite the switch from biological to cultural explanations of human difference (which some suggest followed a move from "overt" to "laissez-faire" racism following the 1960s civil rights movement), both are capable of achieving the same effect in "naturalizing" stereotyping and prejudice. Durrheim and Dixon conclude that lay ontologies are crucial to our understanding of contemporary stereotyping and prejudice, and that the character of these ontologies reveals how the science of human difference remains intimately involved in social projects. They postulate that a scientific discourse with less recourse to "human essence" and greater reference to the impact of historical and contemporary social forces on the production of difference might "provide the basis for a critical [as opposed to an essentialized] common sense."

Section 2: Geneticization and the Nature of Difference

The historical role that social forces have played in the scientific analysis of difference is a theme taken up by the first of four chapters exploring the 'geneticization' of difference. In this chapter, Alice Wexler draws on a detailed critique (Hans and Gilmore, 1969) of two publications by the psychiatrist Percy Vessie (1932; 1939) in which he claimed to have established a link between witchcraft and Huntington's chorea. Hans and

Gilmore (1969) found, among other things, that the hereditary relationships that had been assumed by Vessie and earlier fieldworkers were often erroneous; and acts that were described as beyond the normal range of contemporary behaviors (and therefore deemed symptomatic of witchcraft and/or Huntington's chorea) often were not. Wexler suggests that these mistaken interpretations reveal how Vessie approached his study with very strong preconceptions. She considers this an 'ideological' motivation to link chorea with "the figure of the witch and the hag," a motivation that "led him to the incorrect data and not the reverse." Wexler's interpretation is supported by the fact that Vessie had ready access to information capable of exposing the flaws in his genealogical analyses, yet failed to consult (or heed) these data. However, it is also likely that the prevailing popular and scientific beliefs of the 1930s provided a context in which a biased approach to eugenic research, such as Vessie's, was common if not inevitable.

Wexler tentatively suggests that Vessie, a man of science and of his time, was clearly situated in what one might call a "network of genetics, eugenics and sterilization." His results came out in a way that fit that network and reinforced/reproduced this. As such, Vessie would have been encouraged to co-opt "the twentieth century metaphorical use of the term witch . . . " and use this to reinterpret past events from a contemporary perspective when "witch" meant something entirely different. Wexler suggests that such an approach to "science," and the lack of rigor (deliberate or otherwise) in researching the genealogies of the "witches" and the families concerned, is probably unsurprising within the context of 1930s eugenics. Yet she argues that this example ably demonstrates the way in which scientific and popular conceptions of disease interact with each other to substantiate, and make real, social representations of difference. Wexler concludes that understanding the historical development of such narratives (both popular and scientific) may "help us to become more sensitive to unstated but stigmatizing cultural and social assumptions" that influence scientific practice and the explanations this provides.

The assumptions underpinning the development of scientific theses is an issue explored from an unusual perspective in the next chapter. Here, Jenny Kitzinger examines the way the media represent and challenge genetic research into human difference, drawing on scientific claims for an association between a genetic variant (Xq28) and 'homosexuality' (Hamer et al., 1993), the so-called 'gay gene.' Kitzinger points out that any research into human difference raises questions about how 'difference' is conceptualized by the scientists involved. For homosexuals, such research "enters an explosive area of . . . personal biographies, policy geographies, and political landscape," because concerns about "causation are deeply implicated in personal biographies for many lesbians and gay men . . . "

and reflect the objectivization of sexuality as difference (and homosexuality as different). Claims relating to Xq28 occurred against a backdrop of debate, both among and beyond lesbians and gays, regarding the nature of homosexuality and how this might best be (re)presented to ensure equal rights. Although some prefer 'social' to 'biological' explanations, others question why society is "so concerned with seeing homosexuality as a condition that has causes . . . [and as] a medical or psychiatric [category]" rather than a social one that requires no 'explanation.' From this perspective it is, perhaps, unsurprising that any media reports of the Xq28 findings were likely to be pilloried for promoting both homophobia and genetic determinism. Yet much of the criticism of the media was unwarranted, unfair, and counterproductive.

Looking beyond the headlines (which were both sensational and provocative) and the letters pages (where a good deal of homophobic correspondence was published), Kitzinger found the majority of press reports to be remarkably balanced, and in many respects less strident than promotional copy in, and from, scientific journals. This included: extensive critiques and firm rebuttals of biological determinism; substantial contributions from gay men and activists (both as sources and as authors); considered debate on the "morality, validity, utility and politics of the search for a 'cause' of homosexuality;" the (re-)introduction of important social considerations into the debate (such as prenatal testing and selective abortion); and not-so-gentle lampooning of the research process. Indeed, Kitzinger suggests that an all-too-ready tendency to blame the media for the way in which research on human difference is presented (rather than conducted) is not only inaccurate, but deeply problematic. This is because it preserves the fiction that scientific research is conceived and conducted in a social vacuum (where the "search for truth" is unaffected by social context) and thereby fails to question why such research is undertaken in the first place (i.e., why it is deemed necessary, relevant, or meaningful).

The role of social factors in the selection of topics relevant for research is a theme taken up by Tom Shakespeare, whose chapter sets out to explore the potential impact of genetic research into impairment and illness on social, familial, and individual experiences of disability. Shakespeare looks beyond contemporary fears surrounding neo-eugenicist applications of the "new genetics" (such as the prenatal diagnosis and selective termination of "disabled fetuses") to explore the consequences of such technology on "people who are born." Shakespeare argues that these consequences include the power genetics has in the "naming and knowing" of disability. Identifying those genetic traits associated with, and those responsible for, "disabilities" is likely to reinforce the "medical model" of disability, in which disablement is viewed as an individual affliction rather than an experience of phenotypic difference imposed by mainstream "able-bodied" cultures. This is a view of disability that activists have challenged as victim-blaming, pointing out that disablement is the product of the barriers society throws up to human difference which thereby create the experience of being *dis*-able*d* (in the same way people are gender-*ed* and race-*d* by sexism and racism). Furthermore, Shakespeare argues that the geneticization of disability will undermine the likelihood that society will accept responsibility for these barriers, inasmuch as the cause of disability will be understood to be innate (i.e., biological and genetic) rather than extrinsic (i.e., social and cultural).

That said, Shakespeare argues that the availability of prenatal diagnosis for those genetic conditions predisposing to, or responsible for, "disabilities" will also project responsibility and blame onto parents who "choose" to continue with a pregnancy despite a "positive" prenatal test, and those who do not ("want" or "choose" to) undergo prenatal testing. Although the notion of such "elective disability" might "make the reproductive choices of everyone less free," its impact on the reproductive freedoms of disabled people is likely to be acute. At the same time, the application of genetics in the identification of individuals with a genetic predisposition towards late-onset illnesses has the potential to, as Shakespeare puts it, "turn healthy people into [latently] disabled people," albeit disabled in the sense of "people who face discrimination and prejudice" as a result of their phenotype or genotype and who therefore "find themselves having common cause with disabled people." Shakespeare concludes that it is crucial to maintain a balanced approach to difference, which recognizes that it is the consequence of social structures and social practices that are ultimately responsible for its most profound consequences.

Richard Ashcroft rounds off this section on geneticization by reviewing what he describes as the "cloudy" and changing terrain that links genetic research and medical practice to ideas about race and ethnicity. Ashcroft believes that the vast majority of geneticists recognize that race and ethnicity capture little of the variance in genetic diversity. Yet casual observations linking genetic traits to racial and ethnic groups remain commonplace not only among publics but also in genetics primers and textbooks. This is all the more remarkable when one considers how "painfully aware" most geneticists seem to be of the historical and political baggage of "racialized" approaches to population genetics. Ashcroft argues that there is a clear moral and ethical distinction between research that aims to demonstrate a genetic basis for different racial or ethnic groups, and research that aims to determine group differences in the risk of disease-related genetic traits. However, both are undermined by our inability to reliably operationalize group categories. Indeed, this is only possible when the social relations responsible for group membership are biological (e.g., relatedness) or when group membership is subsequently "biologized" by (re)defining membership with reference to biological criteria (however fortuitous or arbitrary these might be). Thus, although clinicians and geneticists seem to agree that "little sense can be made at the molecular level of the idea of race," Ashcroft argues that race and ethnicity are likely to become more important for diagnosis and treatment until such time as genomic screening becomes routinely available in clinical practice.

Ashcroft offers three reasons why there might be a resurgence of what has become known as "racial profiling" in medicine. (i) Although crude and imprecise, racial/ethnic categorization is accepted practice for disaggregating populations in genetic and related biomedical research, and for classifying individual patients in clinical practice. As a result, data produced by racialized genetic research can be readily translated into racialized clinical practice. (ii) More accurate (genotypic) categories for genetically disaggregating populations are new and expensive, and therefore will only become available and economic for the development of new diagnostic or therapeutic technologies. For this reason race and ethnicity are likely to remain the only available classification schemes for applying past research on existing technologies. (iii) Until such time as research participants and patients are routinely genotyped, clinicians are likely to use race or ethnicity to "play safe" when applying technologies known to have differential side-effects in different racial or ethnic groups. In this way racialized practice may be justified in terms of safety as well as the evidence available on efficacy and cost. Geneticists largely accept that genetically distinct human races do not exist, however, they nonetheless recognize that "an instrumental epidemiological notion of race is both useful and important . . . for practical purposes. . . ." Under these circumstances, Ashcroft concludes that "the very least we can do is to: challenge the social constructions which underpin structural inequities in our society; insist on justification and explicitness; be alert to social practices disguised as neutral measurement; and resist the shift from probability thinking concerning genes to categorical practice concerning races."

Section 3: Scientific Practice and the Pursuit of Difference

Following on from Ashcroft's analysis, Simon Outram and George Ellison go on to explore the scientific practices that sustain the use of race and ethnicity in genetic research in the first chapter of the third and final section. Drawing on interviews with geneticists working on the editorial boards of high-impact genetics journals, Outram and Ellison found that few seemed to recognize that race and ethnicity are fluid and contingent concepts, and are therefore inherently unreliable as scientific classifications. This seemed remarkable given the lack of consensus on how either should be defined or operationalized. There was more consensus on the limited accuracy of race and ethnicity as markers for genetic variation and the interviewees were all aware that race and ethnicity are sensitive concepts capable of generating substantial public controversy. Although they were keen to avoid such controversy, few were prepared to forgo the use of racial or ethnic categories in genetic research, at least not until there was more "conclusive" evidence of their questionable utility. Instead they sought to distance their work from the wider social context in which this took place and attributed the controversy to the public misunderstanding of science.

Outram and Ellison found that these distancing strategies proved to be essentially untenable because contemporary norms of acceptable scientific practice require that race and ethnicity are operationalized using socially salient and socially sanctioned classifications. Indeed the editors they talked to adopted what appeared to be an entirely pragmatic approach in which the social attributes of race and ethnicity were viewed not simply as potential problems but also as potential solutions. This was evident in discussions about: the choice of nomenclature and terminology; the use of social determinants and correlates of group identity to improve the acuity of genetic research; and the impact of social policy on research practice. Thus interviewees were able to circumvent, if not embrace, the view that race and ethnicity are socially constructed, while continuing to use these social categories as potentially useful proxies for the modest genetic variation resulting from the geographical and social characteristics that informed their construction. In the process, Outram and Ellison suggest that geneticists have been able to co-opt many of the social characteristics of race to substantiate its use as a marker of genetic variation. These are characteristics that, somewhat paradoxically, form an integral part of the critique which describes race as "socially constructed." It will therefore be necessary to rearticulate this critique if it is to succeed in preventing the (further) geneticization of race, ethnicity, and unrelated forms of social identity (including those explored by Dingwall et al., Wexler, Kitzinger, and Shakespeare).

The role that genetic research has played in highlighting the treatment of ostensibly "vulnerable" research populations forms an important part of the backdrop to the next chapter by Ricardo Ventura Santos. In the absence of statutes controlling the use of genetic material collected from indigenous communities, the genetic (and commercial) exploitation of samples collected without specific consent (particularly by foreign scientists) has led to calls for the samples to be returned and a growing reluctance to engage in such research in the future. In this chapter, Santos explores the challenges facing research among indigenous and isolated populations, using as an example the differential treatment of Brazilian Amerindians in legislation from colonial times to the present day. From the outset of European colonization, the indigenous population of the Americas was viewed (and/or represented) as inferior: smaller, feebler, and more fragile. These seventeenth- and eighteenth-century views, augmented by the racial science of the nineteenth century, remained influential well into the twentieth century. Following the "conquest" of the Americas, relations between Indians and Europeans developed into an ostensibly benign if explicitly paternalistic approach that sought to protect the rights of the indigenous population much as if they were children.

Thus, just as adolescents aged 16 to 21 were deemed legally "incompetent relative to certain acts" under the Brazilian Civil Code of 1916, so were "forest dwellers." Likewise the 1973 Indian Statute assumed that Indians lacked legal competence and the principle of "guardianship" remained central to Brazilian policies even though the radical Federal Constitution of 1988 moved away from the integrationist and assimilationist policies of the past, and recognized Indian cultural practices as equivalent and legitimate. Nonetheless the more enlightened principles enshrined in the 1988 Constitution are borne out in the Ministry of Health's 1996 "Guidelines and Norms Regulating Research Involving Human Subjects." This guidance excludes Indians from those groups deemed to lack the "capacity of understanding needed for adequate consent" but includes them as the only population classified as a "special thematic area." In part, this reflects the continuing "protection" afforded these groups, in part the drive for greater community participation in the development and delivery of health services and related research (as a result of which indigenous communities are now involved in scrutinizing research proposals). This approach places particular emphasis on the community's needs and on protecting biological and cultural diversity. In the process it clearly favors applied research. In contrast, research into aspects of biocultural difference that appear to have less immediate relevance (including research into genetic diversity) faces substantial challenges, despite the desire among Indian communities for "people outside to learn about our lives." Santos acknowledges that this is partly a consequence of inappropriate and unethical research practices in the past. However, it also illustrates how the resilience of attitudes that sought to protect rather than empower vulnerable communities can undermine the research required to document and preserve their biosocial integrity.

Bioarchaeologists and skeletal biologists face similar challenges, as outlined in the penultimate chapter in this volume by Bethany Turner, Diana Smay-Toebbe, and George Armelagos. However, what is different for these researchers (and others who work on human remains) is that they are unable to consult with their study subjects/participants because, of course, they are dead. This has made bioarchaeology particularly vulnerable to questionable research practices because the subjects/participants are unable to provide consent (or protest) and are unable to inform (or challenge) the analyses conducted. These practices include the unethical collection and retention of skeletal remains from burial sites, and analyses that go beyond the data available to make biased, inaccurate, and exaggerated ancestral claims. The latter include the identification of essentially unidentifiable human remains and the use of discredited racial classifications. Nonetheless, Turner et al. argue that the science of human remains is uniquely valuable because it "enriches what is known of . . . [the] heritage and antiquity . . ." of disadvantaged individuals who were overlooked when they were alive. However, because this work often involves exhuming bodies and handling human remains it is important that the descendants of these groups see this work as "useful and respectful. . . ."

The involvement of indigenous "descent communities" in such work became mandatory in the United States following NAGPRA (the Native American Graves Protection and Repatriation Act). Under NAGPRA, all federally funded institutions were required to compile inventories of any human remains or cultural artifacts they held, and to make these publicly available and subject to repatriation and reburial where appropriate. The legislation caused concern among those who saw valuable material being placed off limits and lost to science. However, others acknowledge that NAGPRA has helped to tackle "an unfortunate legacy of institutionalized, unethical behavior and scientifically justified discrimination," by recognizing "the rights of Native Americans [and others] to decide what becomes of the remains and grave accompaniments of their ancestors. . . ." Moreover, Turner et al. argue that NAGPRA, and related legislation elsewhere, is more than "a long overdue codification of indigenous rights . . . ." The Act has also had a number of (unexpected) benefits for skeletal research such as the standardization of data collection and data coding in osteological databases, and improved access to, and increasing use of, long-neglected and under-researched skeletal collections. Indeed, some Native American communities have appeared willing to permit scientific analyses of skeletal remains now that their rights over these have been recognized by NAGPRA, and what Turner et al. call a "substantial portion of Native Americans have expressed the belief that archaeology and bioarchaeology are beneficial to the preservation of Native American culture." Notwithstanding the concerns raised by Santos in the previous chapter, legislation intended to protect the rights and well-being of research subjects/participants can clearly help to improve opportunities for, and the quality of, research into human biology and the nature of difference.

The tangible benefits of research that adopts a participatory, reflective, and interdisciplinary approach are much of what encourage Alan Goodman, in his concluding chapter to this volume, to suggest a crucial role for anthropology. Drawing on the examples of 'stature' and 'race,' Goodman demonstrates how an interdisciplinary approach can help to reveal the role that culture plays in the development of human biology, both as a subject of enquiry and as a discipline in its own right. In particular, he explains how the social and cultural significance of short stature remains (as a marker of poverty or low social status) even in contexts where short stature is more likely to be genetically prescribed. Thus, in the context of twenty-first-century societies looking to be "better than well," the enduring sociocultural link between stature and status has led to the medicalization of idiopathic shortness. In part this has been driven by commercial pressures, particularly in identifying a market for synthetic human growth hormone (HGH). In the process, as Goodman explains: "a cure of dubious efficacy [HGH] has manufactured a disease." Yet the most powerful insight this analysis provides is how, "In a paradoxical flip of biosocial causality, the biological phenotype of height might still equate itself with wealth if only those with wealth can afford HGH!"

In a similar fashion, Goodman explains how discredited sociocultural ideas surrounding the notion of "race" "became a powerful ideological tool" for explaining and justifying biological difference, even when this is clearly the consequence of social practices. Indeed, because race provides only a weak proxy for genetically determined biological difference, the reasons why it continues to be used as a genetic construct in science and society must be sociocultural. These are: first, because it is useful for those with power (to maintain the status quo); second, because science is essentially conservative (and therefore slow to change); and third, because race is visible in almost everything we do (and therefore difficult to ignore). Goodman argues that those who oppose the use of race as a genetic construct have not come to understand the depths to which this is networked into science and society. With this in mind Goodman concludes by arguing that anthropology is ideally placed to "take the lead in closing the destructive chasm between science and humanity." He offers a compelling argument for retaining an interdisciplinary approach within "four-field anthropology" because, "Without a grounding in the social sciences, biological anthropology is at risk of being just a lesser field of biology." Thus, "In an era of ever increasing specialization, biological anthropology . . . can offer something uniquely valuable: . . . a natural born interdisciplinary field . . . [providing] a deeper analysis of human development, adaptation, variation and behavior." It is a fitting conclusion to the volume, calling upon all who conduct research into human biology to recognize the role of sociocultural forces in the construction of difference.

Tautology, Biology, and Phenomenology

Beyond their collective focus on the role of science, society, and human biology in conceptualizing the nature of difference, the chapters in this volume raise three additional issues with wider ramifications for scientific research into human biology and beyond. The first of these involves the tautological tendencies of all scientific enquiries. The second concerns the exceptional power of biological explanations for 'difference.' The third considers what science and society might do to escape the clutches of biological determinism. These issues suggest that 'naturalization' offers a unifying "analytical framework" that defines difference in a tautological fashion and thereby sustains the resilience of discredited claims. Naturalization involves ascribing differences to natural causes (Foucault, 1978; Kaufman and Morgan, 2005) within an analytical framework that determines how such differences should be defined and understood. Language plays a special role in this process: first as a biological mechanism linking the material, philosophical, and symbolic aspects of human nature;* and second as the medium that facilitates analytical debate yet constrains this within semantic networks of meaning. Despite these constraints, the contributions to this volume indicate that three strategies can successfully challenge the naturalization of difference and expose the role of social forces in the production, recognition, and interpretation of difference. These involve: (i) reflective practice, to address the potential for bias; (ii) phenomenology, to generate alternative analytical frameworks; and (iii) interdisciplinary dialogue, to draw on contrasting perspectives. These strategies allow research into human biology to transcend the tendency towards naturalization and instead offer unique insights into the social construction of difference.

The Role of Analytical Frameworks in the Tautological Tendencies of Science

It seems self-evident that the way we see the world is determined by the way we look at it, by what Ashcroft calls "the tools we bring to the work." The challenge that scientists face is not so much to recognize when their analyses are influenced by a particular analytical framework, but to accept this as an inevitable consequence of their work (Van Fraassen, 1980). This is particularly challenging for those scientists who have been conditioned to assume a mantle of objectivity. Because they deliberately avoid explicitly subjective perspectives, the analytical frameworks that influence what they see tend to be well hidden from view, at least to them. Under such circumstances it is easy to assume that similarities are invisible unless you look for them, and that differences become visible even when you don't look at them. Yet what passes for similarity or difference is determined

* By "material, philosophical, and symbolic" I mean to distinguish between human bodies (our temporal, material selves); human thoughts (our spiritual, metaphysical selves); and the way human bodies and human thoughts are used to represent specific ideas.

by how these are defined (Thiher 1997). Likewise, the visibility of similarity or difference is determined by how these are sought. In this way scientific enquiry into the nature of difference is prone to a perverse tautology: the only differences that can be found are those that match the way difference is defined; and, because these differences match the way difference is defined, they inevitably confirm that the definition works (M'Charek, 2000).

This would not necessarily pose a problem if we had a firm grasp on the definitions and perspectives that scientists use. But even within carefully objectivized scientific practice these often remain undeclared. Instead they take the form of assumptions that have been overlooked or taken for granted as part and parcel of routine practice. As such they make the link between the way we look at the world and what we end up seeing even more difficult to detect. This is doubly important given that what remains unseen remains unexplored and is thereby essentialized as an integral property of the phenomena being examined (Whyte, 2003). Much of this volume is concerned with such unseen links between frames of reference and scientific practice. In particular, many of the chapters explore the way in which social prejudice (in the broadest possible sense) influences the way that science recognizes and interprets difference. Indeed, as we have seen, this is a theme that runs through many of the chapters in this volume (see, e.g., Marks, Kitzinger, Shakespeare, or Turner et al.).

Although links that are unexplored tend to be essentialized, those that are explored tend to become validated. This is because, by studying those issues society deems relevant for enquiry, and by doing so using concepts and techniques that are socially determined, research can create the impression that these issues are 'valid': indeed, 'valid' in both senses of the word: first, that the issues exist; and second, that it is correct to study them in that particular way (even at the expense of other issues and other approaches). In so doing research can create the data required to prove that the issues are not only real (after all, here we are measuring them), but that they are also neutral topics of enquiry (as opposed to loaded artifactual phenomena). This is an important part of the tautological nature of contemporary research (Pool, 1997). What is defined as relevant to study is then made out to be relevant by studying and measuring it.

Take, for example, the availability of data on maternal weight gain during pregnancy.* These have fueled a huge literature on its potential role as a determinant of maternal and neonatal health (e.g., Ellison and Harris, 2002). Yet this research largely ignores a number of careful studies demonstrating that maternal weight gain has little meaning and little utility as a marker for pregnancy outcomes (e.g., Dawes and Grudzinskas, 1991). In effect, a whole subdiscipline, or at least a dedicated line of enquiry,

* These are data that are still collected for largely historical reasons.

can evolve as an artifact of data produced for other, unrelated, or arbitrary reasons. Data on race and ethnicity offer a parallel example. These data are often routinely collected to monitor social injustice (see Ashcroft), yet they then become available for use by others postulating very different explanations for difference, be they scientists (Outram and Ellison) or the lay public (Durrheim and Dixon). Given the pressures on contemporary academic communities (Hvalkof and Escobar, 1998), it is perhaps unsurprising that the availability of data would steer scientific activity in perverse directions (Whyte, 2003). And this is, perhaps, one of the more obvious mechanisms through which analytical frameworks are reproduced: by analyses that rely on data that are produced according to these frameworks.

The impact of analytical frameworks on scientific practice is nothing new to those contributors who adopt perspectives from the humanities (such as Wexler or Ashcroft) or the social sciences (such as Dingwall et al., Durrheim and Dixon, or Shakespeare). In the humanities and social sciences science can be as much an object of study as the natural and social world itself (Williams et al., 2003). To those on the humanities side of what some have called the "two culture divide" (Goodman et al., 2003) understanding (what passes for) material differences in human biology requires a grasp of their wider philosophical and symbolic meanings and the way these meanings reflect and reinforce scientific and social practices. As Kitzinger writes in her chapter on the "gay gene," "Concepts of nature are always used in the performance of culture."

Just as analytical frameworks influence analytical practices elsewhere (Van Fraassen, 1980), so these less tangible properties of the human body inculcate research into human biology with tautological frameworks that are cultural, semiotic, and entirely artifactual. In the process, the human body offers up caricatures and symbols that reflect and impose a preconceived interpretation of our biological nature: the characterization of our "biological nature" is preconditioned by sociocultural processes. These processes are then subject to (re)interpretation in the light of what we subsequently "discover" about our biology through the lens of prior expectation. What is unique for research into human biology is, however, that this circular process reinforces and strengthens a view of human bodies as crafted by (and therefore subject to) the "laws of nature" (Lange, 2000). Indeed the central role that "nature" plays in the popular and scientific conceptualization of biology makes human biological research intensely susceptible to the tautological tendencies common to all science.

The Role of 'Naturalization' as a Unifying Analytical Framework

Our preoccupation with the natural origins of human biology leads us to approach the study of difference in ways that pre-empt alternative explanations. As such, research into human biology exhibits a preference for exploring certain types of difference in certain types of ways: emphasizing differences between populations over differences between individuals;* focusing on biological consequences of natural or physical (as opposed to social) environments; and displaying a preference for genetic over environmental explanations for difference. In both the problems it traditionally identifies and the approaches it adopts to examine these problems, research into human biology tends to emphasize and reify a naturalistic view of human nature at the expense of all other interpretations. This is evident not least in Dingwall et al.'s treatise on criminality, in Marks' analysis of human–ape comparisons, and in Shakespeare's critique of the "medical model" for disability. Much of what passes for scientific enquiry into human biology has become essentially corrupted by a circular argument in which preconceived notions of what biological differences *are* influence which of these are studied and how these are studied. This inevitably results in scientific activity and scientific interpretations that reflect and validate the world view that led us down that line of enquiry in the first place. Although these processes are not unique, as we have seen, to research into human biology it is, perhaps, more acutely felt here than in research on topics less close to home. This is because it is an area of research that is capable of reinforcing our prejudices about the role of biology in human nature, be it a genetic link between witchcraft and Huntington's chorea (Wexler) or the anatomical inferiority of native Americans (Santos and Turner et al.).

However, what is not clear is why, at least in the West, it has proved popular if not compelling to view nearly all forms of difference—including those in behavior, perception, thought, and belief—as essentially biological or, to use today's most fashionable idiom, "in the genes" (e.g., Adam, 2005). Although this may be a phenomenon peculiar to contemporary Western culture, it is nonetheless pervasive (e.g., Pinker, 2002). This seems to hark back to a less rational era when our biological nature was imbued with quasi-metaphysical properties. Biological differences were viewed as the outward stigmata of underlying biological processes that were responsible for "innate" differences in character and temperament, such as criminality (see Dingwall et al.) or even witchcraft (see Wexler). This remains evident today in the extraordinary power of genetic idioms as popular representations and explanations for difference (e.g., Spector, 2003), and the way genetic research has been applied to social issues such as sexuality (see Kitzinger), disability (see Shakespeare), and "race" (see Ashcroft and Outram and Ellison). But what this is also likely to

* Particularly in the way in which genetic variation is attributed to populations on the basis of shared sociogeographical ancestry (Ellison and Jones, 2002).

reflect is the search for a unifying principle for all aspects of human nature grounded in the material rather than the metaphysical. Given our biological origins, where better to look for the material origins of human nature?

This is very much the approach adopted by E.O. Wilson (1988). He famously argued that all things human, be they material or symbolic, scientific or artistic, might ultimately be reduced to the complex biochemical interplay between genes and environments through a process of analytical synthesis he dubbed "consilience".* On the other hand, Stephen Jay Gould (2003), in his last great treatise, concluded that science and the humanities were only unified by their common quest for knowledge. Although he argued that each could benefit from closer collaboration with the other,** he felt they were intrinsically different modes of thought whose synthesis seemed improbable, not least because he had found it impossible to achieve.

Of course both of these views may be essentially correct and neither is incompatible with the other. Metaphysics aside, Wilson's argument seems compelling. Given the computational power necessary, we should be able to comprehensively understand, if not necessarily predict, the interplay between genes and environments that leads to individual differences in human form and function (i.e., our bodies, our behaviors, and our thoughts). However, reducing the world to the nth degree might ultimately prove an unsatisfying exercise for explaining why any particular combination of components came together to produce any particular human being, human action, or human thought (Dover, 2000). Likewise, the principal attraction of Gould's argument (that science and the humanities are qualitatively different and *that's just the way things are*), is simply that it seems to reflect the way we experience these two very different forms of knowledge and knowledge production. Yet this is also unsatisfying because it offers no mechanism for unifying knowledge generated by each approach save a rather wistful appeal to mutual respect and collaboration. Indeed, both Wilson's (1988) and Gould's (2003) attempts to bridge the gap between the sciences and the humanities—both of which are, after all, mechanisms and products of human enquiry—tend towards rather precarious extremes.

As we have seen, in the first chapter in this volume, Cartmill offers an attractive alternative. He argues that language provides the link between our material and philosophical natures, and thereby a link between the sciences and the humanities. Linguistic processes rely on

* After William Whewell, who first coined the term in 1840 (albeit to mean something subtly different; Gould, 2003).

** In what he called, a "consilience of equal regard," with deliberate reference to both Wilson and Whewell.

biological mechanisms yet are also, somewhat paradoxically, capable of releasing human discourse from the bonds of 'nature.' In effect, language is situated within our biological heritage but sets us free to make what we will of our biological and social worlds. This is a satisfying thesis for those struggling to reconcile the material and philosophical aspects of human nature (Blakey, 1998; Ramachandran, 2003). But it is troubling for those of us who, as scientists, want to impose some objective order on the world and set this apart from what passes for 'knowledge' in the arts and humanities. Nonetheless, even for us scientists it should be an appealing perspective, not least because it helps us get a firmer understanding of why different disciplines often examine the world in idiosyncratic ways and why we struggle to communicate or explain our more philosophical and symbolic experiences (Lange, 2000). The principal unifying quality of language is therefore that it provides us with a common universal medium which enables us to articulate and (re)negotiate our definitions of what is biological and what is social, what is fact and what is fiction, what is truth and what is right—an ability that provides the subject matter for Durrheim and Dixon's discourse analyses. The great benefit of Cartmill's argument is that it stems from a concrete material ability linking the biological nature of humans (and all the biochemical interactions between genes and environments which that involves) with our ability to interpret the world as it is and the world as it could (or should) be.

The Role of Language and Semantic Networks in Analytical Frameworks

Certainly it seems clear that we cannot study even the most concrete phenomena without a word to signify or symbolize what it is that we are studying (Ramachandran, 2003). However, it is less obvious that even the most ephemeral and imaginary concepts become available for study, and subject to reification, once we have a name for them. Thus "social capital," currently so much in vogue among those looking for a nonmaterial explanation for the tangible value of human relations, only became available for study when the term achieved widespread currency.* Likewise, the concept of "race," which receives so much attention from contributors to this volume (Marks, Durrheim and Dixon, Ashcroft, Outram and Ellison, Turner et al., and Goodman), is that most slippery of words. It is one that

* First through the initial definition provided by Lyda Judson Hanifan (1916) and then through its subsequent application as a focus for research and policy. Hanifan used the term "social capital" to describe "those tangible substances [that] count for most in the daily lives of people" (1916: 130). The wider application of "social capital" in research and policy is largely attributed to Robert Putnam (1993), who drew on the work of Jacobs (1961), Bourdieu (1983), and Coleman (1988), among others.

conjures up a largely specious interpretation of human difference (both social and biological). It then imposes this as a mechanism, explanation, and justification for essentially artifactual phenomena and concepts: in this case, that biological differences between "racial" groups are natural (Kelleher and Hall, 2003), and that "racial" groups are genetically determined subspecies (Goodman, 2000).

Without language it seems inconceivable that humans could generate a coherent view of their material, philosophical, and symbolic worlds (or at least attempt to do so; Ramachandran, 2003). Yet through language these worlds become intrinsically linked, the one influencing the other. Indeed, perhaps the most important caveat to what we might call the "unification of knowledge through language" is that so much of the exploration and interpretation of these different worlds appears subconscious (Van Fraassen, 1980). This is determined not by discussion but by the structures that govern, constrain, and channel our essentially silent thoughts and analytical practices. Yet this may well be the ultimate strength of Cartmill's argument, because it is only when such links remain *un*spoken, *un*discussed, and thereby *un*examined that artifactual processes are so readily essentialized. In the absence of discussion such issues become not simply tacit but also implicit and integral to the characteristics involved (Thiher 1997).*

This tacit reproduction of ideas and concepts is evident in semantic networks. Semantic networks link words together to create conceptual maps of extended and implied meaning. These maps help us to define and anchor the words we use. In this way the term 'difference' conjures up a whole range of related characteristics that reflect social and cultural investments in beliefs about its nature and meaning. In particular that 'difference' is, in some sense: abnormal, deviant or alien; subject to disdain, fear, or suspicion; and a consequence of some event, process, or circumstance responsible for eliciting variation from the 'norm' or an 'other.' Likewise, as we have already seen, the term 'biological' is linked to a semantic network of terms that are primarily associated with 'nature' both as an essential property of all things, as a product of 'natural selection,' and as the ecosystem that surrounds and engulfs us and of which we are a part.

Thus 'biological' is both an integral and essential part of who we are and an implicit link with our more primitive, subconscious, and presocial origins, when we were just another (natural) product of our planet's extraordinary ecological and evolutionary history. It is therefore hardly

---

* Yet, as the chapters in this volume attest, these issues can be articulated and thereby recognized for what they are—artifacts of the way in which society influences the world in which we live and the way we explore and understand it.

surprising that research into human biology, which focuses on identifying and evaluating the meaning of 'biological difference,' approaches this subject from a perspective tainted by all of the implicit meanings attributed thereto. And in this way language operates as a mechanism for the scientific tautologies we explored earlier. Language achieves this by restricting discussion to those phenomena we have named. Language then situates named phenomena within a semantic network that makes it difficult to separate words from the phenomena and related concepts they come to represent.

The Role of Analytical Frameworks in the Resilience of Discredited Claims

There are two important consequences of the way that scientific tautologies and the vagaries of language tend to naturalize human biological and social differences. These can generate findings that set the science up to ridicule (e.g., Baker, 2001). On the other hand these can make it incredibly difficult to present alternative points of view. There are many good examples of the former among the contributions to this volume. For example, this is evident in Dingwall et al.'s description of Lombroso's (1911) ludicrous caricatures of criminal stigmata; in Wexler's careful critique of Vessie's (1932) biased analyses of historical reports and genealogical data; in Kitzinger's description of the scorn heaped by the media on scientific claims of a 'gay gene;' and in Turner et al.'s summary of the contentious debates surrounding "Kennewick man." Yet each of these contributions also acknowledges that such claims often remain largely intact long after they have been discredited or deemed ludicrous. Although the resilience of discredited claims is disheartening it is not entirely surprising given the tenacity of scientific paradigms (as Outram and Ellison describe). They are also situated within (or re-enter) a world that is preconditioned to receive them, one in which they make intuitive sense. In part this is because claims embraced by lay audiences tend to be retained long after they have gone out of scientific fashion. In part this is simply because the world today is still structured in a similar way to the world that existed when these ideas first originated. Thus the resilience of discredited claims simply reflects the conservation of social structures, attitudes, and beliefs that produced and subsequently sustained such claims. Why else would researchers continue to examine the XYY syndrome as a potential determinant of criminality (Gotz et al., 1999)? Why else would Vessie's (1932) story continue to be cited as relevant to our understanding of Huntington's chorea (Aronson, 2002)? And why else would the biological basis of homosexuality continue to be a focus of enquiry (e.g., Camperio-Ciani et al., 2004)?

The social inertia underpinning the resilience of discredited claims might also explain why alternative ideas and novel perspectives often struggle for acceptance. These not only run contrary to accepted principles but also tend to adopt ways of seeing that can seem heretical to the social structures and mores that determine what can be seen and how this should be interpreted. Again, 'race' provides a salutary example of a discredited concept that nonetheless reflects a view of the world that has proved difficult to address from alternative perspectives (Duster, 2003). The tendency of "race" to invoke innate biological explanations for differences that are externally produced has even encouraged some to lobby against its use as an analytical concept (e.g., Ellison et al., 1996; Martin, 2005; Ashcroft). But this disregards, if not denies, the *lived experience* of anyone whose social identity has been "raced" (West, 1993). Quite simply, their lives cannot be adequately understood without reference to the role that discredited theories of racial difference play. Yet this, more subjective, approach seems anathema to the way in which variables are defined and measured in "science" as reliable and universally valid concepts. It requires us to accept and embrace the personal and contingent nature of experiences we cannot objectify or generalize (Millward and Kelly, 2003; Phoenix, 2004).

Finally, it seems likely that the resilience of discredited claims also indicates that these were inadequately or perhaps inexpertly discredited. Perhaps the crucial critiques were poorly worded, translated, or disseminated, or were inordinately complex and inaccessible to those they most needed to reach. If so, this is yet another role that language plays in the analytical frameworks which generate scientific tautologies and the naturalization of difference. In this, language operates not simply as a mechanism for the exchange of information, thoughts, and beliefs—something that analytical frameworks (and related 'theories') thereby facilitate—but also as a device for separating and isolating cultures and communities of practice.* As such it is crucial to address the barrier of language when exposing the tautologies that underpin the naturalization of difference. For these critiques to take hold they must, like any other argument, adopt language that is accessible and persuasive to audiences from the natural and social sciences, arts and humanities, and beyond. This is a tall order, and an activity few academics embrace, not least because it warrants little acclaim from those whose opinions they value most, academic colleagues within their own discipline. Indeed a crucial place to start is within the academy itself by adopting not simply a composite multidisciplinary discourse but a truly cooperative interdisciplinary dialogue (Hvalkof and

* This is an issue that Dingwall et al. and Durrheim and Dixon explore, and which Ashcroft cites as a potential flaw in the apparent "translatability" of racialized findings from the genetics laboratory to the clinic.

Escobar, 1998). This might help achieve what Gould (2003) called a "consilience of equal regard."

The Role of Anthropology in Challenging Analytical Frameworks

It is in this sense that Goodman, in this volume's concluding chapter, argues for a key role for "four-field anthropology"* as an inherently interdisciplinary discipline. Unfortunately it is a discipline that has suffered its fair share of squabbles among subdisciplines and a recent spate of intradisciplinary separation and divorce. Nonetheless, anthropology does seem to offer a rare opportunity for regaining what Dingwall et al. depict as a halcyon age when it was possible for an academic to traverse the natural and social sciences, arts and humanities. This is because, as Goodman describes, some of the most profound insights into the nature of human differences have come from anthropologists** who have managed to draw on the perspectives and methods favored by biological *and* social anthropologists. In the process they demonstrate the interplay between our genetic and cultural heritage, our natural and social environments, and our material, philosophical, and symbolic worlds in the production and understanding of difference. Goodman and Leatherman (1998) use the term "biocultural," to describe this more integrated approach to studying human biology. This approach recognizes that the biological differences we see among living humans represent the consequences of natural and social processes that have become incorporated into our bodies, quite literally "embodied." In this way, our bodies become a "text that is interpreted, inscribed with meaning—indeed made—within social relations" (Thomson, 1997: 22).***

Although anthropology appears to be an ideal context in which the biological and social sciences might be integrated, most biologists and most social scientists would not necessarily identify with anthropology, not least those who left the discipline to avoid the naturalistic pitfalls of biological

---

* "Four-field anthropology" traditionally provides a liberal arts training in four subdisciplines: archaeology, biological/physical anthropology, cultural/social anthropology, and linguistics.

** Including two contributors to this volume: Marks and Armelagos.

*** Indeed, a number of analysts with interests spanning the divide between the natural and social sciences have also been trying to come to terms with the interaction between the biological and the social in the production of difference. And like those who seek to define and embrace a new anthropological synthesis, such analysts have sought to introduce new labels (such as "biosociality," Rabinow, 1996; and "ecosocial," Krieger, 2001) to signify a more interdisciplinary approach—one that explicitly recognizes the interaction between science and society in the production, recognition, and interpretation of "difference."

anthropology and the contemporary challenges facing "pure" anthropological research (see Santos and Goodman). Yet these researchers also have much to gain from forging closer ties with each other. In particular, it would help biologists recognize the human body as not simply a biological product but also as the material substrate that makes social interactions possible. As such it subsequently reflects, both in the embodiment of social circumstances and through the symbolic interpretation of biological features, the social production of biological difference. Effectively, what is recognized as biologically different depends more upon social perception than biological reality. Likewise, issues are often legitimized as biological based on social preferences as opposed to material evidence. Indeed, some might argue that biologists have a moral and ethical responsibility to engage more fully with social scientists to break through the "false epistemological ceiling" that limits their gaze to "natural" processes and "natural" explanations. Without this they will continue to overlook or downplay social forces and misattribute social consequences to natural causes.

The Role of Reflective Practice, Phenomenology, and Interdisciplinary Dialogue in Exposing the Naturalization of Difference

No one would deny that this is likely to prove a difficult and contentious endeavor. But social science perspectives on human biology and scientific practice are crucial. They are necessary to disentangle those instances where it is legitimate to view biological differences as natural from those where they are the consequences of social mechanisms or the result of cultural emphases that have no foundation in what we know (rather than believe) to be true about the 'natural' world. That said, social scientists also have much to learn from biologists about the impact of society on human bodies (e.g., Birke, 2003). Indeed, human biology has much to tell us about the social world because the social world has had such a profound role in the biological differences we see. What these differences are, and how society differentiates between those deemed relevant for study and those that are taken to be 'natural,' tells us reams about social power, privilege, and advantage. For example, this, as Turner et al. emphasize, is one of the great strengths of bioarchaeology in that it offers an insight into the lives of powerless, underprivileged, and disadvantaged peoples who were largely invisible and who history has a tendency to forget. Research into human biology therefore has a crucial role to play in exposing the consequences of what we have previously called "advantage inherited posing as advantage accrued on the basis of equal opportunity and merit" (Ellison and De Wet, 2002: 145). Power, privilege, and advantage have all been directly implicated in the production of biological difference. But what is often less clear is how these benefits are concealed behind analytical frameworks that overlook or misdiagnose difference and invoke alternative, natural, explanations.

The contributions to this volume illustrate how research into human biology can adopt one of three specific strategies to escape from the naturalization of difference and expose the role that social prejudice plays in the construction of difference. The first of these (which is, perhaps, most evident in Cartmill, Marks, Ashcroft, and Turner et al.) involves embracing the reflective practices common to ethnographic research, in which the frameworks, assumptions, and preconceptions of researchers are exposed for consideration as potential biases. The second involves challenging prevailing paradigms using phenomenology to identify alternative frameworks and explanations for difference (and is a particular feature of Durrheim and Dixon, Shakespeare, Outram and Ellison, and Santos). The third involves drawing on perspectives from different disciplines to identify hidden assumptions and adopt a more flexible and contingent view of human nature (an approach that is prominent in Dingwall et al., Kitzinger, Wexler, and Goodman). Whichever technique we adopt, research in human biology has much to add to our understanding of human nature and the nature of difference.

Acknowledgments

This introduction has benefited enormously from the many friends, colleagues, and students who have discussed the issues involved and commented on numerous drafts along the way. Particular thanks to Thea de Wet, Jane Wills, Simon Outram, Mary Halter, Yvonne Erasmus, and Mark Gilthorpe. The finished product also draws extensively on the advice, wisdom, and counsel of my co-editor, Alan Goodman, whose conclusion to this volume aims to counterpoise the more epistemological and philosophical issues addressed here. I trust that his careful editing and gentle coaxing are evident throughout and I hope that these do justice to his efforts to wrest clarity and simplicity from a challenging synthesis of this volume's diverse perspectives on *The Nature of Difference*. Like he, I am grateful to him for giving me the freedom and opportunity to try to say what I wanted to say.

REFERENCES

Adam, D., Female orgasm all in the genes, *Guardian*, June 8, 2005.

Aronson, S., Witchcraft was not sorcery but a too human disease, *The Providence Journal*, July 22, 7, 2002.

Baker, R., *Fragile Science: The Reality Behind the Headlines*, Macmillan, London, 2001.

Birke, L., Shaping biology: Feminism and the ideal of the "biological," in *Debating Biology: Sociological Reflections on Health, Medicine and Society*, Williams, S.J., Birke, L., and Bendelow, G.A., Eds., Routledge, London, 2003.

Blakey, M.L., Beyond European enlightenment: Toward a critical and humanistic human biology, in *Building a New Biocultural Synthesis: Political-Economic Perspectives on Human Biology*, Goodman, A.H. and Leatherman, T.L., Eds., University of Michigan Press, Ann Arbor, MI, 1998.

Bourdieu, P., Forms of capital, in *Handbook of Theory and Research for the Sociology of Education*, Richards, J.C., Ed., Greenwood, New York, 1983.

Camperio-Ciani, A., Corna, F., and Capiluppi, C., Evidence for maternally inherited factors favouring male homosexuality and promoting female fecundity, in *Proceedings of the Royal Society of London Series B—Biological Sciences*, 271, 2217–2221, 2004.

Coleman, J.C., Social capital in the creation of human capital, *American Journal of Sociology*, 94, S95–S120, 1988.

Dawes, M.G. and Grudzinskas, J.G., Repeated measurement of maternal weight during pregnancy—Is this a useful practice? *British Journal of Obstetrics and Gynaecology*, 98, 189–194, 1991.

Dover, G., *Dear Mr Darwin—Letters on the Evolution of Life and Human Nature*, Phoenix, London, 2001.

Duster, T., Buried alive: The concept of race in science, in *Genetic Nature/Culture: Anthropology and Science beyond the Two-culture Divide,* Goodman, A.H., Heat, D., and Lindee, M.S., Eds., University of California Press, Berkeley, CA, 2003.

Ellison, G.T.H. and De Wet, T., "Race", ethnicity and the psychopathology of social identity, in *Psychopathology and Social Prejudice*, Hook, D. and Eagle, G., Eds., Juta, Cape Town, 2002, pp. 139–149.

Ellison, G.T.H. and Harris, H.E., Gestational weight gain and "maternal obesity", *Nutrition Bulletin*, 25, 295–302, 2000.

Ellison, G.T.H. and Jones, I.R., Social identities and the "new genetics": Scientific and social consequences, *Critical Public Health*, 12, 265–282, 2002.

Ellison, G.T.H., De Wet, T., IJsselmuiden, C.B., and Richter, L.M., Desegregating health statistics and health research in South Africa, *South African Medical Journal,* 86, 1257–1262, 1996.

Foucault, M., *The History of Sexuality, Vol. I: An Introduction*, translated by Robert Hurley, Pantheon, New York, 1978.

Gillham, N.W., *Sir Francis Galton: From African Exploration to the Birth of Eugenics*, Oxford University Press, Oxford, 2001.

Goodman, A.H., Why genes don't count (for racial differences in health), *American Journal of Public Health*, 90, 1699–1704, 2000.

Goodman, A.H. and Leatherman, T.L., Eds., *Building a New Biocultural Synthesis: Political-Economic Perspectives on Human Biology*, University of Michigan Press, Ann Arbor, MI, 1998.

Goodman, A.H., Heat, D., and Lindee, M.S., Eds., *Genetic Nature/Culture: Anthropology and Science beyond the Two-culture Divide,* University of California Press, Berkeley, CA, 2003.

Gotz, M.J., Johnstone, E.C., and Ratcliffe, S.G., Criminality and antisocial behaviour in unselected men with sex chromosome abnormalities, *Psychological Medicine*, 29, 953–962, 1999.

Gould, S.J., *The Hedgehog, the Fox, and the Magister's Pox—Mending and Minding the Misconceived Gap between Science and the Humanities*, Jonathan Cape, London, 2003.

Hamer, D., Hu, S., Magnuson, V., et al., A linkage between DNA markers on the X chromosome and male sexual orientation, *Science*, 261, 320–326, 1993.
Hanifan, L.J., The rural school community center, *Annals of the American Academy of Political and Social Science*, 67, 130–138, 1916.
Hans, M.B. and Gilmore, T.H., Huntington's chorea and genealogical credibility, *Journal of Nervous and Mental Diseases*, 148, 5–13, 1969.
Hvalkof, A. and Escobar, A., Nature, political ecology, and social practice: Toward an academic and political agenda, in *Building a New Biocultural Synthesis: Political-Economic Perspectives on Human Biology*, Goodman, A.H., and Leatherman, T.L., Eds., University of Michigan Press, Ann Arbor, MI, 1998.
Jacobs, J., *The Death and Life of Great American Cities*, Random House, New York, 1961.
Kaufman, S.R. and Morgan, L.M., The anthropology of the beginnings and ends of life, *Annual Review of Anthropology*, 34, 317–341, 2005.
Kelleher, D. and Hall, B., Ethnicity and health: Biological and social inheritance, in *Debating Biology: Sociological Reflections on Health, Medicine and Society*, Williams, S.J., Birke, L., and Bendelow, G.A., Eds., Routledge, London, 2003.
Krieger, N., A glossary for social epidemiology, *Journal of Epidemiology and Community Health*, 55, 693–700, 2001.
Lange, M., *Natural Laws in Scientific Practice*, Oxford University Press, Oxford, 2000.
Lombroso, C., Introduction, in *Criminal Man According to the Classification of Cesare Lombroso*, Ferrara, G.L., Ed., Putnam, New York, 1911.
Martin, P., The paradox of race/ethnicity, *Critical Public Health*, 15, 77–78, 2005.
M'charek, A.A., *Technologies of Similarities and Differences: On the Interdependence of Nature and Technology in the Human Genome Diversity Project*, University of Amsterdam, Amsterdam, 2000.
Millward, L.M. and Kelly, M.P., Incorporating the biological: Chronic illness, bodies, selves, and the material world, in *Debating Biology: Sociological Reflections on Health, Medicine and Society*, Williams, S.J., Birke, L., and Bendelow, G.A., Eds., Routledge, London, 2003.
Moscovici, S., The phenomenon of social representations, in *Social Representations*, Moscovici, S. and Farr, R., Eds., Cambridge University Press, Cambridge, 1984.
Phoenix, A., Extolling eclecticism: Language, psychoanalysis and demographic analyses in the study of "race" and racism, in *Researching Race and Racism*, Bulmer, M. and Solomos, J., Eds., Routledge, London, 2004.
Pinker, S., *The Blank Slate*, Penguin, London, 2002.
Pool, R., *Beyond Engineering: How Society Shapes Technology*, Oxford University Press, Oxford, 1997.
Putnam, R. D. *Making Democracy Work. Civic Traditions in Modern Italy*, Princeton University Press, Princeton, NJ, 1993.
Rabinow, P., Artificiality and enlightenment: From sociobiology to biosociality, in *Essays on the Anthropology of Reason*, Rabinow P., Ed., Princeton University Press, Princeton,1996.
Ramachandran, V., *The Emerging Mind*, Profile, London, 2003.
Spector, T., *Your Genes Unzipped*, Robson, London, 2003.
Taylor, L., *Born to Crime: The Genetic Causes of Criminal Behaviour*, Greenwood, Westport, CT, 1984.
Thiher, A., *The Power of Tautology: The Roots of Literary Theory*, Associated University Presses, Cranbury, NJ, 1997.

Thomson, R.G., *Extraordinary Bodies: Figuring Physical Disability in American Culture and Literature*, Columbia University Press, New York, 1997.

Van Fraassen, B.C., *The Scientific Image*, Oxford University Press, Oxford, 1980.

Vessie, P.R., On the transmission of Huntington's chorea for three hundred years: The Bures family group, *Journal of Nervous and Mental Disease*, 76, 553–573,1932.

Vessie, P.R., Hereditary chorea: St. Anthony's dance and witchcraft in colonial Connecticut, *Journal of the Connecticut State Medical Society*, 3, 596–600, 1939.

West, C., *Race Matters*, Beacon, Boston, 1993.

Whyte, J., *Bad Thoughts: A Guide to Clear Thinking*, Corvo, London, 2003.

Williams, S.J., Birke, L., and Bendelow, G.A., Eds., *Debating Biology: Sociological Reflections on Health, Medicine and Society*, Routledge, London, 2003.

Wilson, E. O., *Sociobiology: The New Synthesis*, Harvard University Press, Cambridge, 1975.

PART 1

LOOKING FOR 'NATURE' IN HUMAN NATURE

1

IS THERE A BIOLOGICAL BASIS FOR MORALITY?

Matt Cartmill

CONTENTS

SOCIOBIOLOGY AND MORALITY

The idea that there is a biological basis for morality has acquired a bad odor, arising largely out of the controversy that began with the 1975 publication of E. O. Wilson's book *Sociobiology* (Caplan, 1978; Segerstråle, 2000). Although the ostensible topic of that book was the general biological principles that govern the evolution of social behavior in animals, its real focus was Wilson's wish to find a way of grounding moral judgments in statements about natural selection. The book's moralistic keynote is sounded in its opening sentences:

> Camus said that the only serious philosophical question is suicide. That is wrong even in the strict sense intended. The biologist, who is concerned with questions of physiology and evolutionary history, realizes that self-knowledge is constrained and shaped by the emotional control centers in the hypothalamus and limbic system of the brain. These centers flood our

> consciousness with all the emotions — hate, love, guilt, fear, and others — that are consulted by ethical philosophers who wish to intuit the standards of good and evil. What, we are then compelled to ask, made the hypothalamus and limbic system? They evolved by natural selection. That simple biological statement must be pursued to explain ethics and ethical philosophers . . . at all depths. (Wilson, 1975:3)

Wilson returned to this theme in the book's final chapter, in which he attempted to bring the theory of group selection home to the study of human behavior:

> Scientists and humanists should consider together the possibility that the time has come for ethics to be removed temporarily from the hands of the philosophers and biologicised. . . . Only by interpreting the activity of the emotive centers as a biological adaptation can the meaning of the [moral] canons be deciphered. (Wilson, 1975:563)

Wilson predicted that the sociobiological investigation of morality would be complicated. Divergent selection pressures, he asserted, will have instilled different moral intuitions — what he called "the activity of the emotive centers" — in the brains of different human groups. "Some of the activity," wrote Wilson, "is likely to be outdated, a relic of adjustment to the most primitive form of tribal organization . . . directed toward such Pleistocene exigencies as hunting and gathering and intertribal warfare." Other emotive-center activity "may prove to be *in statu nascendi*, constituting new and quickly changing adaptations to agrarian and urban life" (Wilson, 1975:563, 575). Because the evolutionary interests of men differ from those of women, and those of children, adolescents, and parents differ from each other, we can expect natural selection to yield different moral intuitions in different sex and age groups. Likewise, because patterns of differential reproductive fitness in expanding populations will be different from those in static or shrinking populations, different nations and ethnic groups "will tend to diverge genetically . . . in ethical behaviour." It follows, in Wilson's view, that "no single set of moral standards can be applied to all human populations, let alone all sex-age classes within each population" (1975:563–564). A truly scientific morality will thus enjoin us to treat children differently from adults, men differently from women, and Moldavians differently from Masai.

On the last page of *Sociobiology*, Wilson foresaw a time, probably around the end of the twenty-first century, when our species will have achieved "an ecological steady state." In that day, he predicted, selection pressures will be more nearly uniform across nations and classes, and "a full, neuronal explanation of the human brain" will enable us to identify brain mechanisms underlying our moral intuitions. Guided by evolutionary sociobiology, scientists can then proceed to "mold cultures to fit the requirements of the ecological steady state" and help move us toward "a genetically accurate and hence completely fair code of ethics" (Wilson, 1975:574–575).

This is an unattractive prospect, which dismayed even Wilson. "When we have progressed enough to explain ourselves in these mechanistic terms, and the social sciences come to full flower, the result might be hard to accept," he admitted. "In . . . the ultimate genetic sense, social control would rob man of his humanity" (Wilson, 1975:575). Although Wilson saw this grim technocratic future as the inevitable end state of what he called "our autocatalytic social evolution," he took some consolation from the fact that he, at least, would not live to experience the dreadful fruition of his new science. His book ends with this extraordinary sentence: "But we still have another hundred years" —*Après moi, le deluge.*

Today, with 30 of those hundred years already behind us, it seems unlikely that any of Wilson's predictions will come to pass. Most of the human species is still divisible into rich nations with shrinking populations and impoverished nations with expanding populations. While we know a lot that was not known in 1975 about the workings of the human brain and the structure of the human genome, the possibility of explaining or predicting human behavior in mechanistic terms seems as unattainable as ever. Perhaps most unhappily for Wilson's program, only a small minority of social scientists invoke sociobiological concepts in attempting to account for the phenomena of human minds, societies, and history. Outside the school of evolutionary psychology, the consensus appears to be that it is both vain and vicious to try to explain differences in ethical beliefs with reference to hypothetical differences in allele frequencies and brain structures.

FITNESS VS. GOODNESS

It goes without saying, though it still needs to be said, that there is essentially no evidence for any of the neurological uniformities and differences that some evolutionary psychologists posit to explain variation in human behavior, much less for the imaginary genetic variation that is supposed to underlie the imaginary neurological variation. However, even if we were to accept the fundamental assumptions of this approach — that our genes determine the way our brains are built and that the way our brains are built influences the choices we tend to make — the Wilsonian program still seems to miss the point. The task of moral philosophy is not to determine why we act the way we do, but to decide how we *should* behave. Accounts of the causes of human behavior have only a limited relevance to this job. If we were to discover, for example, that pedophiles are sexually attracted to children because of a mutation affecting some brain structure, that might help us to decide whether pedophilia should be treated as a crime or an illness, but it would not help us to decide whether we should evaluate it as morally good, bad, or indifferent. The decision would not be made any easier if we knew the relative fitness of the mutation. These biological facts are irrelevant to the moral issues.

Wilson's suggestion that ethics can be "biologicised" by "interpreting the activity of the emotive centers as a biological adaptation" thus seems fundamentally misguided. No matter how much we learn about the genetic and neurological bases for human actions and appetites, facts about human DNA sequences and neuronal arrangements cannot furnish us with a moral compass. Even if some human appetite were known to be innate and genetically determined — and even if it were shown to be a product of natural selection (which is not the same thing) — that would not imply that it deserves to be gratified. There is no reason to assume that what natural selection has produced is good. To quote Bertrand Russell (1966:24), "If evolutionary ethics were sound, we ought to be entirely indifferent as to what the course of evolution may be, since whatever it is is thereby proved to be the best."

Political critiques of evolutionary approaches to ethics have long focused on just this point — that such approaches seem to imply that whatever is, is right. As the moral philosopher Abraham Edel put it,

> Any institution claiming roots in instinct can . . . clothe itself with the moral authority of absolute fixity. The history of social psychology is strewn with the wreckage of instincts intended to prop up prevalent institutional forms, such as pugnacity instincts to prop up war, acquisitive instincts to reinforce private property, and a variety of specific instincts to support the family. (Edel, 1955:116)

Russell's dictum exposes the vanity of attempts to ground morality in instinct. We are *not* indifferent to the course of evolution, and everything that natural selection favors is not thereby proved to be right. Therefore, no matter what we may concede about natural selection as an *explanation* for our moral impulses, it will not serve as a *foundation* for ethics. From the standpoint of the moral philosopher, the sociobiological approach represents a sort of explained intuitionism, which finds the source of moral judgments in our appetites and impulses and offers a Darwinian explanation of why those appetites and impulses are as they are and not otherwise. There is nothing inherently suspect about such explanations, and many of them must in fact be true. For example, sugars taste good to us but potassium cyanide does not, because our sense of taste has evolved to promote the ingestion of nourishing substances and discourage the ingestion of poisons. However, such explanations of our appetites cannot serve as a justification for any moral judgment. Adducing them for this purpose is an instance of the notorious error known as the naturalistic fallacy — the logically vain endeavour to sneak the word *ought* into the conclusion of a deductive argument without having introduced it into any of the premises. Hume's injunction against this fallacy deserves quotation:

> In every system of morality, which I have hitherto met with, I have always remark'd, that the author proceeds for some time in the ordinary way of reasoning, and establishes the being of a God, or makes observations concerning human affairs; when of a sudden I am surpriz'd to find, that instead of the usual copulations of propositions, *is*, and *is not*, I meet with no proposition that is not connected with an *ought*, or an *ought not*. This change is imperceptible; but is, however, of the last consequence. For as this *ought*, or *ought not*, expresses some new relation or affirmation, 'tis necessary that it should be observ'd and explain'd; and at the same time that a reason should be given, for what seems altogether inconceivable, how this new relation can be a deduction from others, which are entirely different from it." (Hume, 1740[1898]:245–246).

Sophisticated heirs to the Wilsonian tradition are aware of this fallacy, and they avoid committing it. For example, the evolutionary psychologists Randy Thornhill and Craig Palmer argue that rape may be an innate, adaptively conditioned behavior of human males, fixed by natural selection and driven by a so-called "rape module" hard-wired into the male brain (Thornhill and Palmer, 2000). They indignantly insist that demonstrating the existence of such a brain module, or of a reproductive advantage for the rapist, would not justify the commission of rape. They protest that the critics who attack their ideas as furnishing an excuse for rape are themselves committing the naturalistic fallacy. I think they are right on both counts, but in saying these things, Thornhill and Palmer demonstrate the inadequacy of evolutionary theory as a replacement for moral philosophy. If selective advantage can accrue to wicked impulses as well as virtuous ones — and of course it can — then it cannot furnish criteria for distinguishing between the two.

A moral theory has to provide such criteria. Moral theories are brought into being by reflecting on moral dilemmas, which arise when human impulses are in conflict. The practical job of such a theory is to help us sort out our impulses and decide which ones to obey in a particular situation. Adaptive considerations cannot be relied upon to resolve such conflicts, because there is no law of nature that prevents adaptively useful impulses from conflicting with each other.

Such impulses often do conflict. We find them doing so in our own lives, and this sort of internal conflict is not uniquely human, but for other animals, such dilemmas can only be aesthetic or pragmatic, not moral. This distinction is hard to explain in purely adaptive terms. It is easy enough to dream up plausible-sounding adaptationist explanations for aesthetic or moral preferences alike, but it is not clear how adaptive considerations by themselves can explain why these are two distinct *sorts* of preferences — why, for example, hating oppression and hating cold rain seem like preferences of different kinds, though they may be of equal adaptive importance to a social mammal with no fur. If this is a real difference, and other animals possess aesthetic but not moral discrimination, then the ability to make moral judgments must depend on some distinctively human faculty that we do not share with our animal relatives and that comparative evidence of a Wilsonian sort cannot help us to understand.

All of these considerations justify many of the harsh indictments that the sociobiological and evolutionary–psychological versions of ethics have received from philosophers and social scientists. Wilson's critics are right to conclude that attempts to explain morality in biological terms have thus far proved "irrefutable, useless and dangerous" (Solomon, 1980) and that "all statements about the genetic basis of human social traits are necessarily purely speculative" (Lewontin et al., 1984:251) in the present state of our knowledge. However, "speculative" does not imply "mistaken." Though a moral theory cannot be validly derived from biological facts alone, any workable moral theory has to be partly grounded in assumptions about our biological nature. Unless Wilson is right in some sense — unless there is some innate basis for human social traits and dispositions — compelling moral argument, persuasion, and reform are impossible, because there is no other common ground from which unforced negotiation can proceed.

In what follows, I want to argue five main points:

1. There is little point in debating moral issues unless valid moral arguments are possible.
2. Such arguments are possible only if there are quasi-universal shared values that allow us to judge between conflicting moral principles.
3. Such values must be grounded in some way in human biology. Other possible foundations for moral negotiation are either inadequate to the task or ultimately reducible to claims about our biological nature.

4. Facts about human biology are necessary but insufficient to generate moral principles. To hold otherwise would be to commit the naturalistic fallacy.
5. The *ought* term emerges in the conclusions of moral arguments by virtue of the fact that prescriptive moral judgments are universalisable (Hare, 1963). In essence, this is a fact about the grammatical structure of human language.

MORAL ARGUMENTS AND THEIR PREMISES

The central task that confronts any moral reformer is to get beyond the surfaces of conflicting moral codes and injunctions to discover more deeply held shared premises that can be invoked in evaluating those codes and injunctions themselves. If valid and compelling moral argument is possible, there must be some things that are true of all people everywhere — or at least of most people in most times and places — that have moral implications.

It may be the case that no such arguments are valid, because no such moral quasi-universals exist. Some have argued that no society or ethical tradition has the right to preach to another — that moral arguments cannot penetrate the boundaries between cultures, because different cultures hold and express different and incommensurable values. However, if we accept this sort of relativism, it is not clear why most of the contributors to this book have bothered to write their chapters, for if cultural tradition determines what is right, what grounds can any of us have for criticizing the moral assumptions of our own cultures and societies? It does not serve the cause of moral reform to reject the doctrine that whatever is, is right, only to embrace the idea that whatever people think is just, is just. Both notions are recipes for stasis, not progress. Indeed, it is meaningless to talk of historical progress without assuming the existence of nonarbitrary, morally preferable directions of change, in terms of which progress can be evaluated (Chomsky, 1975).

At its core, then, relativism is not progressive but reactionary. It is also not ethnographically realistic. Like the historical myths of religious fundamentalists, it looks backward to an imagined time when human social units were small, homogeneous, and characterized by what Durkheim (1933) called "mechanical solidarity," but in modern complex societies, there is never a unitary moral consensus that is accepted by all the people who live, work, and trade together on a daily basis. The peaceful operations of such a society require that the conflicts between different moral traditions and assumptions be negotiated. Such negotiations — and the quasi-universal shared values that make them possible — are a necessity of practical politics in our time. If we have learned nothing else from the geopolitics of the opening years of the twenty-first century, we should have learned at least this much.

Assuming that there are quasi-universal shared values that can provide a foundation for moral negotiation, do they necessarily have to be somehow grounded in human biology? I believe that they do, because the other possible foundations for moral argument turn out to be unacceptable, inadequate, or ultimately biological in nature. There are at least six such unsatisfactory possibilities:

i. Some moral principles might not be empirically true but *logically necessary*, as a presupposition of thought, or at least of moral discourse. The rules of modal logic come to mind here — for example, the principle that *must* implies *can*, meaning that one is never obliged to do the impossible. Another example might be Henry Sidgwick's so-called Principle of Fairness, which is the proposition that "individuals in similar conditions should be treated similarly" (Sidgwick, 1890). Rules of this sort may constitute indispensable "boundary conditions" for the formulation of moral norms with practical content (Markl, 1980), but abstract logical principles with no empirical content cannot oblige us to particular concrete actions. Again, Hume has the last word here. "Morals," Hume wrote (1740[1898]:235), "have an influence on the actions and affections. . . . It follows, that they cannot be deriv'd from reason . . . because reason alone can never have any such influence. The rules of morality, therefore, are not conclusions of our reason." Logical necessity thus cannot provide a foundation for moral argument. Hare makes the same point differently:

> It might be said that a principle of conduct was impossible to reject, if it were *self-contradictory* to reject it. But if it is self-contradictory to reject a principle, this can only be because the principle is analytic. But if it is analytic, it cannot have any content; it cannot tell me to do one thing rather than another. (Hare, 1952:41).

ii. A second alternative to grounding morality in biology is the possibility that some moral principles may not be logically necessary, but may be *practically necessary*. It might be that "there are some moral rules that all societies will have in common, because those rules are necessary for society to exist" (Rachels, 1986:531). For example, no moral code that forbids all of its adherents to reproduce can be generally accepted for long, because its adherents will eventually die out. However, an immortal species might not encounter this difficulty. What is practically necessary for us is itself determined by the laws of nature and the facts of human biology. Talking about what constitutes a "practical necessity" for humans is simply another way of talking about human nature.

iii. Many people claim that valid moral principles cannot derive from either abstract logic or the contingencies of human nature but must ultimately be rooted in an authoritative *supernatural source*. Such claims are rarely of practical help in negotiating moral conflicts, since people with different moral codes often have different religious beliefs. However, even if this were not the case, appeals to supernatural authority are still not capable of providing a valid basis for moral argument. As Plato pointed out in the *Euthyphro*, virtue is not good because the gods love it; rather, the gods (if they exist) love virtue because it is good. If moral rules were whatsoever the gods chose to make them, then those rules would be capricious and would cease to have any real moral authority. A grounding of morality in supernatural authority might be salvaged by a natural-law interpretation — for example, by claiming that God has made us so that certain principles of action are right for us, for the same sorts of reasons that reciprocating action in a straight line is right for a saw. This interpretation, though, like the aforementioned argument from practical necessity, again refers the standards of right and wrong to claims about human nature.

iv. One might argue that contingent but nonbiological facts about the universe exist that can provide us with a moral compass. For example, a conventional Marxian or Hegelian might argue that history has an inevitable directionality and that whatever action conduces toward the predetermined end of the historical process is by definition virtuous. This sort of argument seems to me to be subject to the same objections that Russell raised toward evolutionary ethics — namely, that if it were true, we ought to be entirely indifferent to what the course of history may be, since whatever it is is thereby proved to be the best. The laws of thermodynamics imply that entropy increases, but that discovery does not justify the conclusion that the heat death of the universe is a desirable outcome or that we should all try to increase entropy by leaving the lights on at all times. To argue that virtue consists of accelerating inevitable historical processes is just one more form of the naturalistic fallacy.

v. It seems hard to avoid the conclusion that compelling arguments capable of settling moral conflicts must in some way be based on claims about human nature, but this does not necessarily entail claims about human biology. In principle, we might argue that the morally relevant aspects of human nature are not biological.

For example, they might be aspects of our *spiritual* nature, intrinsic properties of the human soul, and, of course, this sort of property is what many people in the past had in mind when they wrote about human nature. Such a claim can evict biology from the moral realm, but only at the price of introducing spiritual substances and nonmaterial modes of inheritance into our picture of the universe. This entails adopting some form of dualism, with all its attendant paradoxes and antinomies. It seems like too big a price to pay for what we get in exchange.

vi. At the London conference that gave rise to this volume, some participants suggested that human nature is socially and historically constructed by human beings themselves — that what is universally and essentially human changes through time and that it does so as part of a historical process that is responsive to the operations of human will and reason. This interpretation seems to me to represent a change in the meaning of the phrase *human nature*. We may of course use the phrase in this way, if we choose, but human nature thus redefined cannot furnish a standard for judging the outcomes of human behavior. If human nature is the product of human choices, then we can choose to alter it, and if we can choose to alter it, then its properties cannot provide a criterion for directing or evaluating our choices. A compass that obligingly points in whatever direction we choose is of little use in navigation.

To sum up, workable moral principles must be at least partly grounded in quasi-universal facts of human nature, and any materialist conception of human nature must be biological, but trying to derive ethics solely from biology leads nowhere. Selectionist explanations cannot justify any moral judgment, since similar explanations can be advanced for conflicting or immoral impulses. To claim that certain actions or principles are good because they are progressive in an evolutionary sense — or because they promote the survival of the human species, which has intrinsic value — is to commit the naturalistic fallacy. Worse yet, Darwinian explanations of morality may be taken as dismissing morality as a natural but objectively baseless prejudice for the same reason that naturalistic explanations of religious beliefs are perceived as a dismissal of religion.

GRAMMAR AND MORAL UNIVERSALS

How can the moral theorist escape from this dilemma? If ethics must be in some way grounded in biology, but mere facts about our biology have no prescriptive force, then where does the prescriptive force come from?

I am not a moral philosopher, but I would like very briefly to sketch what seems to me to be the right kind of answer, since it also has an anthropological dimension. That answer lies in the universalisation of moral judgments that is inherent in the nature of language. I suggest that morality, like most other uniquely human traits, is an epiphenomenon of language and that prescriptive force derives from the way predicates and pronouns work in all linguistic systems.

I follow Hare in thinking that moral judgments are not simply prescriptive, but also have "descriptive meaning." Describing something as "good" or "just," like describing it as "brittle" or "luminous," commits us to the proposition that anything exactly like the subject of the judgment in the relevant respects, including context, possesses the same property. Therefore, unlike other sorts of prescriptions (such as: "Look out!" or "Shut the door!"), moral judgments are *universalisable*. As a result, they are not only prescriptive but are also capable of being utilized in logical constructions. As Hare puts it,

> [Any] moral judgement has to be made with some general moral principle in mind, and its purpose must be either to invoke that general principle or to point to an instance of its application . . .. It is not always easy to elicit just what the principle is; but it always makes sense to ask what it is. The speaker cannot deny that there is any such principle. The same point might be put another way by saying that if we make a particular moral judgement, we can always be asked to support it by reasons: the reasons consist in the general principles under which the moral judgment is to be subsumed. (Hare, 1952:176)

Many people have argued plausibly that human self-awareness is also an epiphenomenon of language. To quote the linguist Derek Bickerton,

> What seems to us most striking about our kind of consciousness is its self-reflexive nature. We can perform a series of actions and at the same time observe ourselves performing them, so to speak . . .. This feeling of subjective consciousness is, of course, "What it is like to be a human", and language contributes to it in a variety of ways. The most basic of these lies in providing the infrastructure for consciousness. You can't look at the spot you're standing on now if there is nowhere else for you to stand. A minimal prerequisite for self-consciousness is a place . . . from which a part of you can look at another part of you. The secondary representational system [of language] is such a place." (Bickerton, 1990:208–209).

In effect, language gives me a virtual place outside my own mind and body where I can stand and make third-person judgments about myself, simply by swapping around pronouns and putting myself in another person's place. I suggest that this same sort of operation lies at the heart of moral judgments. If I perceive that John is contemptible, repulsive, cruel, or wicked because he does X, that constitutes a *prima facie* reason for my not doing X. Putting yourself in the other person's place is at heart a grammatical operation. As such, it is a human universal, which underlies basic principles of moral argument found throughout recorded history, from the "golden rules" of Confucius, Jesus, and Maimonides to Kant's Categorical Imperative and Sidgwick's Principle of Fairness. All these principles are grounded in the same universal properties of language that make us self-aware and that underlie the familiar reproof of a parent to a naughty child: "How would you like it if somebody did that to *you*?"

Moral judgments, then, can be described as the expression of personal feelings tempered by the demands of descriptive universalisation and logical consistency. If so, then any language-using animal would be expected to make moral judgments. To the extent that such judgments are universalisable, their content must ultimately be grounded in quasi-universals of human nature. If morality does not have such a foundation — which, for materialists, means a biological basis — then we cannot sort out our internal and external conflicts by invoking moral principles.

Some people think principles are unnecessary. Pragmatists like Richard Rorty hold that moral truth is whatever wins in a free and open exchange of views. However, in the absence of shared premises, I see no way of telling who wins in such an exchange, apart from the ultimate pragmatic recourse of waiting to see who gets control of the organs of state force. Moral particularists like Jonathan Dancy (2000) contend that invoking overriding moral principles always does violence to the subtleties of individual cases. The problem with this is that any reason for thinking that a principle that applies in one case will not do in another case is itself an overriding moral principle.

As far as I can see, we need shared principles to have any hope of negotiating moral differences without recourse to force. The only reliable and noncontingent foundations for such principles appear to be the operations of logic and language on the one hand and the facts of human biology on the other. Both are necessary to support moral judgments. If the foundations of morality are not to be sought in these things, then compelling moral argument is impossible.

It may be that compelling moral argument *is* impossible. Two arguments to that effect strike me as deserving special attention. First, human individuals differ biologically. If moral arguments must be grounded in our biological nature, are we obliged to tailor our moral decisions to the variable biological properties of the individuals involved? I think that we are but that this is not particularly problematic. We already do this sort of thing all the time, and it is reasonable that we should do so. For example, people whose genetic makeup makes it impossible for them to acquire language in any form (including signing or writing) cannot be condemned for refusing to testify in a court of law. There is nothing intrinsically defective about a moral principle that applies only under certain specified circumstances and to individuals with certain specified properties — provided that the circumstances and the properties specified are in fact relevant to the rights, obligations, and treatments being prescribed (Rachels, 1990). The important fact to bear in mind here is that (contrary to what Wilson hints) there appear to be no morally significant biological properties possessed by some human populations but not others. Moldavians and Masai have different moral codes, but this is a contingency of cultural history. There is no genetic difference between these two ethnic groups that obliges or licenses us to hold them to different standards of conduct.

A potentially more serious objection derives from the possibility of altering human nature by genetic modification. We are able to do this in certain limited ways already, and there is little doubt that we will be able to effect more extensive and fundamental modifications in the future. However — as already noted — if human nature becomes the product of human choices, its properties can no longer furnish criteria for directing or evaluating those choices. In deciding how and whether to alter our own biology, we cannot make use of guidelines and principles derived from the current nature of our species. Clearly, much of the anxiety surrounding the burgeoning technology of genetic engineering stems from a perception of this fact. Such anxiety underscores the importance of assumptions about human nature in moral argument. When such assumptions can no longer be made because human nature has become the product of human choices, the foundations of moral argument will revert to the naked logical and grammatical principles that underlie moral reasoning. Whether these will be sufficient to guide our choices in modifying our own biology remains to be seen.

REFERENCES

Bickerton, D., *Language and Species*, University of Chicago Press, Chicago, 1990.

Caplan, A. L., Ed., *The Sociobiology Debate*, Harper & Row, New York, 1978.

Chomsky, N., *Reflections on Language*, Pantheon, New York, 1975.

Dancy, J., The particularist's progress, In *Moral Particularism*, Hooker, B. and Little, M., Eds., Oxford University Press, Oxford, 2000, pp. 130–156.

Durkheim, E., *The Division of Labor in Society*, Macmillan, New York, 1933.

Edel, A., *Ethical Judgment: The Use of Science in Ethics*, Macmillan, New York, 1955.

Hare, R.M., *The Language of Morals*, Oxford University Press, Oxford, 1952.
Hare, R.M., *Freedom and Reason*, Oxford University Press, Oxford, 1963.
Hume, D., *A Treatise of Human Nature, Being an Attempt to Introduce the Experimental Method of Reasoning into Moral Subjects; and Dialogues Concerning Natural Religion,* Vol. 2, Longmans, Green and Co., London, 1740 [1898].
Lewontin, R.C., Rose, S., and Kamin, L., *Not in our Genes: Biology, Ideology, and Human Nature*, Pantheon, New York, 1984.
Markl, H.S., Reports on discussions: Group One, In *Morality as a Biological Phenomenon: The Presuppositions of Sociobiological Research*, Stent, G.S., Ed., rev. ed., University of California Press, Berkeley, 1980, pp. 209–230.
Rachels, J., *The Elements of Moral Philosophy*, McGraw-Hill, New York, 1986.
Rachels, J., *Created from Animals: The Moral Implications of Darwinism*, Oxford University Press, Oxford, 1990.
Russell, B.A.W., *Philosophical Essays*, rev. ed., Allen & Unwin, London, 1966.
Segerstråle, U., *Defenders of the Truth: The Battle for Science in the Sociobiology Debate and Beyond*, Oxford University Press, Oxford, 2000.
Sidgwick, H., *The Methods of Ethics*, Macmillan, New York, 1890.
Solomon, R.C., Reports on discussions: Group Three, In *Morality as a Biological Phenomenon: The Presuppositions of Sociobiological Research*, Stent, G.S., Ed., rev. ed., University of California Press, Berkeley, 1980, pp. 253–274.
Thornhill, R. and Palmer, C.T., *A Natural History of Rape: Biological Bases of Sexual Coercion*, MIT Press, Cambridge, 2000.
Wilson, E.O., *Sociobiology: The New Synthesis*, Harvard University Press, Cambridge, 1975.

2

BIOLOGICAL DETERMINISM AND ITS CRITICS: SOME LESSONS FROM HISTORY

Robert Dingwall, Brigitte Nerlich, and Samantha Hillyard

CONTENTS

INTRODUCTION

The debate between the biological and social sciences over the explanation of human behavior is at least two thousand years old. Its outlines can be discerned in the arguments between Aristotelians and Stoics over the nature of human nature and the extent to which this is preformed at birth or constructed through interactions between newborns and their social environments. While it is important not to retroject our current concepts into that debate in an anachronistic fashion, we should not underestimate the sophistication that both brought to the argument. The Aristotelians acknowledged the potential influences of inheritance and uterine experience, while the Stoics explored the notion of the innate capacity of infants to interact with and be influenced by their environment. Their model did not presume a blank page so much as a page already prepared to be written upon. Taking a global sweep over the intellectual history of Western Europe and those parts of the planet where its models of knowledge have been imposed or otherwise come to dominate, we can see that neither viewpoint has been able to secure a conclusive victory in the argument. At various times and places, one or the other has secured a local dominance. Explanation of both the victories and their transience presents interesting challenges for the sociology of knowledge, which are not directly taken up here. This chapter simply aims to restate the social scientists' intellectual case against those who argue that biology plays a large part in the direction of human behavior. We do not consider the ethical objections to biological determinism and its social implications. For these to come into play, we need first to be persuaded that the biologists' case has any substantial intellectual legitimacy.

This chapter is organized in three main sections. The first reviews the attempt to produce a biological explanation of crime and the intellectual problems that many social scientists would immediately see with this enterprise. The second looks more deeply at the source of these problems, particularly in early twentieth century debates between social and biological scientists about the nature and role of instinct. It will be proposed that *instinct* played a comparable explanatory role to *gene* in contemporary arguments. Finally, we shall consider what would be required of biology to produce explanations that would carry conviction among social scientists. In order to keep within the editors' prescribed length, we are going to talk very loosely about biologists versus social scientists. The group of scholars considered here would include a variety of people, some of whom might more naturally be described as psychologists or physical anthropologists. Equally, we do not mean to imply that everyone with a degree in biology thinks in the way that we are critiquing. We simply need a shorthand description for those natural/biological scientists who think that much, if not all, human conduct can be explained in terms of biological structures or processes.

THE "BORN CRIMINAL"

The foundational text of biological approaches to crime is Lombroso's *L'Uomo Delinquente*, first published in 1876. This formed part of a wider movement, following in the wake of Darwin, to explore the possible applications of evolutionary thought to human behavior. The socially disadvantaged were seen as those who were less suited biologically to the conditions of late nineteenth century society. Their wretched condition was not the result of social inequality or economic exploitation but of their lesser fitness to compete. Lombroso originally attributed criminality to atavism, a notion derived from Darwin (1981:i, 173), who suggested that the unpredictable appearance of "some of the worst dispositions . . . [may be] reversions to a savage state." The atavistic criminal was a being "who reproduces in his [sic] person the ferocious instincts of primitive humanity and the inferior animals" (Lombroso, 1911: xiv). They could be recognized by a variety of physical characteristics that made them look more like an ape or an ancient or inferior race of humans. These included heavy jaws, prominent brow ridges, large ears, abnormal dentition and so on. Interestingly, from the outset, Lombroso also confused biological and social markers by listing tattoos among these physical stigmata. Although the evolutionary argument might have some consistency in relation to the rest of the list, tattoos are not, of course, heritable in a strictly Darwinian sense.

By the fifth (1897) edition of his book, Lombroso had both introduced other types of criminal and qualified the original thesis. He added the "epileptic criminal" and the "insane criminal" and modified his strong claims about atavism to acknowledge a large fringe of occasional criminals, who might show weak physical signs and could only be precipitated into crime by specific environmental influences. Lombroso subsequently extended this thesis to female offenders who proved to be considerably less distinct as a type from other women, although generally tending toward more virile features (Lombroso and Ferrero, 1895).

Although Lombroso's work was fairly rapidly discarded in its specific detail, the core idea of a biological basis for criminality continued to have a strong influence. For example, a line of work runs through Kretschmer (1921), Sheldon (1940), and the Gluecks (1950, 1956) and was still being pursued during the 1960s by Conrad (1963), looking at body shapes and criminality. Broadly speaking, mesomorphs, with hard, rounded bodies, were said to be more likely to be criminals and ectomorphs, with thin fragile bodies, less likely. Conrad's work suggested that body shape changed from mesomorphic to ectomorphic as children grew up, so that criminals represented a lower level of "ontogenetic development." If one sees this in the context of the discredited but remarkably persistent claim by the nineteenth century biologist Ernst Haeckel that ontogeny recapitulates phylogeny (Gould, 1977), then one might reasonably ask whether Conrad is simply rephrasing Lombroso's atavism.

More recent work has moved from gross physical markers to the molecular level, reflecting a general shift in the focus of biologists' work (Kay, 1993). In the 1960s, for example, there was interest in the possibility that chromosomal abnormalities might have some relationship to criminality. A number of studies found higher incidences of men with an extra Y chromosome among some specialized incarcerated populations (see the summary in Taylor et al., 1973). However, this no longer seems to be

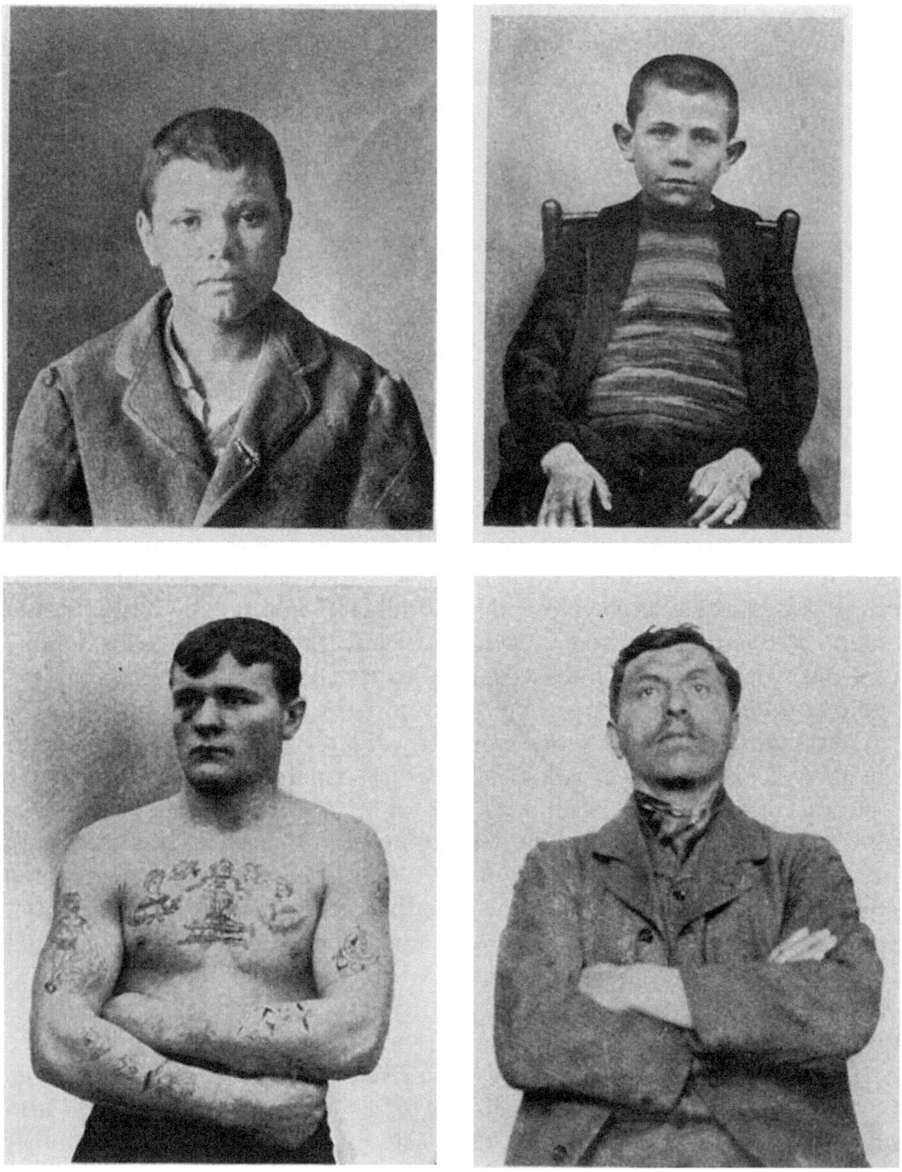

Figure 2.1 Illustrations taken from Lombroso's (1911) *Criminal Man*. Clockwise from top left: "Boy Morally Insane;" "Boy Morally Insane;" "Head of criminal;" and "Anton Otto Krauser Apache." Lombroso famously described tattooing as often revealing: "obscenity, vindictiveness, cupidity..." (1911:232).

fashionable: a recent text by Rowe (2002), one of the leading biologists working on criminality, makes no reference to this body of work. The research frontier has moved from the chromosomal to the genomic level. Brunner et al. (1993), for example, identified a region of the X chromosome that normally codes for an enzyme called monoamine oxidase A (MAOA), which is responsible for breaking down the neurotransmitters serotonin and norepinephrine. In some individuals, a mutation in this gene meant that no MAOA was produced. Men with this mutation seemed to be particularly prone to violence, and the finding was widely reported as the discovery of a "gene for crime." In practice, biologists are generally somewhat more cautious than journalists about making such claims (see Kitzinger, this volume). What Plomin (1994) has called the OGOD (one gene, one disease) relationship is acknowledged to be rare, particularly in the context of complex polygenic traits like criminality. Nevertheless, one can question whether the acknowledgment is sufficient to produce methodologically robust conclusions about the contribution of genetics to behavior, even if it is accepted that the contribution reflects the interacting effects of numerous genes operating in quite sophisticated ways with very specific environmental conditions.

An example of current leading-edge biological work in this field would be a recent paper in *Science* (Caspi et al., 2002), which was also predictably covered as the discovery of an OGOD gene for crime. What the paper actually claims is that childhood maltreatment interacted with a functional polymorphism in the gene encoding for MAOA, leading to a systematic variation in the level of antisocial problems caused by these individuals. A New Zealand birth cohort of 1037 children (52% male) had been tracked from birth to age 26. Maltreatment before the age of 11 was defined by a combination of behavioral observations, parental reports, and retrospective reports by study members at age 26. Antisocial problems were defined by psychiatric assessments for "conduct disorder" according to DSM-IV at age 11, 13, 15, and 18; court records of violent convictions in adulthood; the MPQ aggression scale; and symptoms of antisocial personality disorder reported by close associates of the study participants. Maltreatment was associated with antisocial problems. However, the combination of maltreatment with a low MAOA genotype increased the risk of antisocial outcomes, while maltreatment with a high MAOA genotype did not. It is argued that this association may be consistent with a causal relationship between stressful experiences and antisocial behavior mediated by neurotransmitter development.

Caspi et al.'s (2002) paper is elegantly constructed within its own paradigm assumptions and reasonably judicious in its conclusions. However, many social scientists would find it entirely unconvincing for the same reasons that they found Lombroso's arguments unconvincing more than a century ago. These reflect a very different way of understanding the environment and, ultimately, of understanding behavior. Social scientists are consistently fascinated by the way in which methodological rigor deserts biologists when they are faced with social environments. It is, perhaps, telling that social scientists normally refer to environments in the plural, where biologists refer to environment in the singular. It is as if a physiological, biochemical, or molecular focus reduces everything outside to an undifferentiated blur, the world seen by the myopic without their distance reading glasses.

Lombroso's physical anthropology was derived mainly from measurements of incarcerated populations. The problem with such populations is that they are not a straightforward sample of those who commit criminal acts. Incarceration is the result of a long series of screening decisions, beginning with the victim of a crime, or the first person to respond to that event, and their decision whether to seek the intervention of the criminal justice system or not. At each stage, from investigation through arrest, charge, trial, and conviction, the population is winnowed in non-random ways that lead to the particular pattern of variables that characterize the incarcerated. It is, then, important to distinguish between those factors that are associated with the original act and those that are associated with decisions to screen in or out at each stage of the criminal justice process. Lombroso's error was to argue that a set of physical characteristics that led men and women to be regarded as particularly brutish and threatening to respectable society were related to their acts rather than to law enforcers' screening decisions. The same point was made in relation to sex chromosome abnormalities, which often led to distinctive physiques:

> Even if their behavior was no more aggressive that XXY males, it might be that because of their great height and build they would present such a frightening picture that the courts and psychiatrists would be biased to direct them to special hospitals for community safety. The bias might be further aggravated by the associated intellectual abnormality. This factor might find expression in the raised incidence of XYY (and XXYY) males in special hospital groups. (Hunter, 1966:984)

The question that we might reasonably ask about Caspi et al.'s (2002) paper, then, is whether the factors that lead both professional and lay reviewers of the subjects' behavior to define them as in some sense disturbed also lead criminal justice system personnel to process them in ways that increase their risk of conviction and incarceration. It is not necessarily the case that they are any more or less antisocial but that they are more likely to be selected for high tariff processing.

However, this argument has already begged an even more fundamental question, namely, that of determining what counts as antisocial behavior. The Caspi team conflates judgments of adolescent conduct disorder ("adolescents displaying a persistent pattern of behavior that violates the rights of others"), convictions for violent crimes (common assault, aggravated assault, domestic violence, manslaughter, rape), self-reports on an aggression scale, and ratings by associates. At one end, we have a very precise statement of a direct relationship between genotype and MAOA production. At the other, we have a very diffuse set of indicators of some sort of troublesomeness. Antisocial behavior, however, depends on what counts as pro-social behavior in particular cultural environments. Rowe (2002:3) sneers at "some social deconstructionists [who] say that crime is an entirely arbitrary cultural invention" and goes on to assert that murder and adultery are universally prohibited. Unfortunately, he is simply wrong. For example, as the essays in Bohannan (1960) show, the intricate ways that traditional African societies classify deaths as homicide or suicide does not necessarily map onto contemporary European conceptions that rest on a complex jurisprudence that is used to define the difference between murder, manslaughter, and accidental death. Murder may be universally prohibited, but what constitutes murder is highly contingent on the processes that are used in particular societies to define the significance of acts. The precision at a micro-scale is not matched by an equal precision at a macro-scale. Essentially the same points can be made about adultery and Rowe's other examples (stealing food and telling untruths), all of which are the contextual product of social processes by which behaviors are defined (rather than Platonic ideals with meanings inherent in the acts).

If these acts had the same thing-like quality as MAOA, then it might be easier to sustain the claim for a causal association. Unless a comparable behavioral object, independent of social definitions, can be specified, however, the claim makes very little sense. The best that seems possible is that some people with low levels of MAOA seem to cause other people some unspecified trouble, while others with high levels of MAOA seem not to. Even then, we may still need to know more about what *trouble* means. Like a lot of this type of research, the Caspi et al. (2002) study appears to invoke a standard of gentility, which may reflect the world as the authors would like it to be but which may not have much to do with the world as it is outside the groves of academe. The person who does not show guilt after doing something bad, for instance would have been a hero to the existentialist philosophers of the 1940s and 1950s. A central theme in Sartre's work is the irrelevant and unproductive nature of guilt as an influence on our actions. The person who is impulsive, rushes into things without thinking, may have an important role on the sports field or in armed combat. A soccer team, like an army, may need both thinkers and doers — people who will take risks without evaluating them.

The same arguments lead to comparable problems at the other end of the process. Caspi et al. (2002) fall into the same error by treating maltreatment as equally homogenous. They run together markers for a lack of parental affection, neglect, severe physical punishment, and unwanted sexual contact. Interestingly, the last of these is reported as being derived from self-reports of any genital touching before the age of 11, as well as grosser acts of attempted or actual intercourse. Again, there is a certain prissiness and insensitivity to cultural variation. Consider this account from seventeenth century France:

> The modern reader of the diary in which Henri IV's physician, Heroard, recorded the details of the young Louis XIII's life is astonished by the liberties which people took with children . . .. Louis XIII was not yet one year old: "He laughed uproariously when his nanny waggled his cock with her fingers." An amusing trick which the child soon copied . . .. When he was just over a year old he was engaged to the Infanta of Spain; his attendants explained to him what this meant, and he understood them fairly well. They asked him: 'Where is the Infanta's darling?' He put his hand on his cock . . . . (Ariés, 1973:98–100)

Ariés continues for two pages, concluding with Louis's defloration of his bride at the age of 14. As he stresses, such behavior is very strange to modern sensibilities. However, on evolutionary timescales, it is simply not credible to suppose that the *sequelae* of genital touching have changed so much in 400 years. As pointed out in a review written 20 years ago, the definition of child maltreatment is highly malleable depending upon the picture that investigators wish to present (Graham et al., 1985). The prevalence and incidence of maltreatment can be constructed more or less at will (Dingwall, 1989).

The core of the problem lies in the biologists' philosophical realism, a position that they share with most practicing natural and physical scientists. Realism does not produce many problems in the everyday conduct of natural science. Any competent geneticist can see the existence of the polymorphism, or at least regard the inference of its existence from indirect measures as unproblematic. The association of the polymorphism with varying levels of MAOA rests upon a chain of intervening processes whose activities are clearly recognized, if not always fully understood. That understanding is, however, merely a matter of time as the process of research leads the collective knowledge of the community of natural scientists into closer correspondence with the reality that it is observing.

The sociology of science's skepticism about this epistemology is almost incomprehensible to most working natural scientists. When you hold an Eppendorf tube up to the light and see a tangle of DNA at the bottom, you are seeing something that seems very real and nonarbitrary. Social scientists work with very different materials that make straightforward realist positions simply unsustainable — which is not, of course, to say that there have not been periodic attempts to sustain them. The problem with the biological explanation of crime is that it attempts to cross from one kind of object to another without recognizing the need to confront the epistemological challenges that arise in the process.

The difficulties involved have been well recognized by social scientists since the differentiation of biology, psychology, and sociology was made between 1880 and 1920. At the beginning of this period, a man like Herbert Spencer could write authoritatively about all three. By the end, they were institutionally distinct disciplines, with their own research agendas, journals, and networks of support and patronage. In the process of moving from the undifferentiated homogeneity of the sciences that had characterized scholarship from the revolution of the seventeenth century to the distinct heterogeneity that we recognize in contemporary scholarship, however, there was a willingness to engage in direct arguments that has since largely disappeared, for pragmatic (if not entirely good) institutional reasons (Abbott, 2001).

PHYSIOLOGICAL PSYCHOLOGY AND SOCIAL PSYCHOLOGY

Within 20 years of its first publication, the line of reasoning suggested by Lombroso's work was under sustained attack, particularly by a generation of Americans whose graduate studies in Germany had been cross-fertilized by the proposals about the nature of mind and behavior emerging from pragmatism, their country's unique contribution to philosophy. W.I. Thomas (1896), for example, discusses the limited progress made by "psycho-physics" and the discrediting of Lombroso's work by Baer's measurements of "normal" individuals. Thomas rejected the attempt to ground psychology in biological structures, such as brain weight or cranial measurement. However, he retained a notion of drive or instinct, particularly in relation to food and sex — although the expression of these drives was environmentally determined:

> It is a popular view that moral and cultural views and interests have superseded our animal instincts; but the cultural period is only a span in comparison with prehistoric times and the prehuman period of life, and it seems probable that types of psychic reaction were once for all developed and fixed; and while objects of attention and interest in different historical

> periods are different, we shall never get far away from the original types of stimulus and response. It is indeed a condition of normal life that we should not get too far away from them. (Thomas, 1901:751)

For a period, there was a flurry of interest in trying to define human instincts as the drivers for behavior. McDougall (1909), for example, lists flight, repulsion, curiosity, pugnacity, subjection, self-assertion, the parental instinct, reproduction, gregariousness, acquisition, and construction. Thomas elaborated his "food and sex" instincts into his legendary "four wishes": recognition, response, new experience, and security. By 1921, Faris was pointing to the dire confusion that had resulted:

> How does it happen that gifted men are so unable to agree on what they consider the basic facts of human nature? Some slight differences might be understood, but surely the range is distressingly wide. One [instinct] or two, or four, or eleven, or sixteen, or thirty, or forty — this looks suspicious. (Faris, 1921:188)

Faris notes that instinct has usually been explained by "the so-called genetic method" (1921:198). By this he means a Lamarckian process, where previously advantageous behaviors are impressed on the human organism in an enduring fashion. Two examples that he takes from a contemporary psychologist are the suggestion that the love of baseball reflects prehistoric man's need to run, throw, and strike, while the former dependence of humans on horses is shown by the desire of children to ride rocking horses. If one is talking about societal evolution, Lamarckianism is a more viable theory than in the case of biology: clearly social groups can study their competitors and seek to incorporate their behavior. However, the idea that this is then somehow fixed into physical human structures has all the problems that Darwin and Wallace identified. Several millennia of circumcision have not, for example, led to Jewish boys being born without foreskins. One could respecify the theory in more Darwinian terms. For example, boys who are successful at running, throwing, and striking are advantaged in mate selection and have passed on skills that are then transferable to baseball. However, as Faris notes, such arguments are quickly falsified by ethnology. If we know that the human species has only a minor and relatively trivial degree of genetic variation, then either this gives rise to a high degree of uniformity of behavior or it has very little influence at all. Why would the same selection process result in baseball in the United States, cricket in India, and pelota in the Basque regions of Spain and France? Moreover, Faris points out, our account of the selection process is entirely mythical. McDougall (1909), for example, describes the "primitive family" in terms that he borrows from the folklorist Andrew Lang:

> The primitive society was a polygamous family consisting of a patriarch, his wives and children. The young males, as they became full-grown, were driven out of the community by the patriarch who was jealous of all possible rivals to his marital privileges. They formed semi-independent bands hanging, perhaps, on the skirts of the family circle, from which they were jealously excluded. From time to time the young males would be brought by their sex impulse into deadly strife with the patriarch, and, when one of them succeeded in overcoming him, this one would take his place and rule in his stead. (McDougall, 1909:282)

No one has ever observed such a society. This is simply a "just-so" story. Faris goes on to present a very entertaining account of his six-month-old baby's "instinct for toe-sucking" in terms of its advantage in recycling food dropped on cave floors! Frequently, he adds, these just-so stories are also supported by highly selective examples from lower animals. "Such naïve inventions based on a theory of evolution," he concludes, "form no part of a valid scientific method" (Faris, 1921:193).

Faris acknowledges the role of instinct in animals and possibly in respect of simple acts by very young children. However, he sees no conclusive evidence that humans have any specific instinctive patterns. The 'genetic psychologist' assumes that which he or she should make a hypothesis. An instinct must be capable of universal expression. Ethnology, or as we would now say social anthropology, consistently falsifies any such claims. However, Faris does open an interesting possibility, namely, the study of temperament. Where instinct deals with humans in the aggregate, temperament would deal with them on the basis of individual differences. He insists that temperament is as much a hypothesis as instinct, but that it may be more profitable to pursue, even if only because it has received less investigation, at least as of 1921. We shall return to this suggestion.

Although an interest in biology continued among sociologists for a while after Faris' attack, there is no doubt that this interface went into a decline from which it has never fully recovered. A good index of this can be seen in the historiography of writing about George Herbert Mead, who has probably had the most enduring influence of the scholars working at the boundary between philosophy, psychology, and sociology before World War I. Mead himself had a considerable interest in the embodiment of humans. His first book (Mead, 2001), apparently intended for publication in 1910 but never returned to the printer, devotes roughly a third of its length to a discussion of the field of social psychology and its relationship to physiological psychology. This remains a recurrent theme of the lecture course on *Mind, Self and Society* published by his students in 1934 after his death (Mead, 1962). However, Mead's leading interpreters, Herbert Blumer (1969) and Anselm Strauss (1977), both discarded this dimension of his work. The recent rediscovery of the body as a topic in sociology has rarely led back to this agenda but has, rather, been caught up with the postmodern turn in microsociology that treats the body as a cultural artifact rather than as a topic in its own right. The core of Mead's social psychology is his explication of the basis on which acts acquire meaning.

Mead (1962) begins from a critique of JB Watson's (1925) behaviorism and Darwin's (1872) writing on emotions. Both, he argues, have misconceived the relationship between physical states and behavior in humans by overgeneralizing from studies of lower animals. Lower animals communicate in an automatic fashion by means of gestures and responses. Two dogs seeking to establish which is dominant will run through a fixed sequence of behaviors culminating in the withdrawal of one or the other, through an equally predictable display. Human communication is, however, selective and symbolic. We do not have an undifferentiated response to environmental stimuli. Mead refers to the emerging literature on the psychology of attention to support this claim.* Our responses to our environments are selected and organized through our ability to use symbols. Unlike animal communication, symbols provide for intervening processes between gesture and response and for the entry of the social into these processes. The most important symbols are those of language, which is a shared and collective experience: as Wittgenstein (1972) later emphasized, the notion of a private language is simply nonsensical. Language is intersubjective or it is nothing. The particular mental processes that Mead proposed may no longer justify much discussion. In many ways they are as much a just-so story as the better-known Freudian trinity of ego, superego, and id. However, his analysis of the centrality of language remains central. Because we cannot know what is in another person's mind, we can only infer this from their behavior and from the observation of their response to our inferences. The meaning of our actions is not to be found in our intentions — which are inaccessible — but in others' responses.

The development of conversation analysis since the 1960s, with the help of modern recording technologies, provides an empirical demonstration of what Mead could only contend, namely that, at its simplest, all

* Although the nature of the text does not lend itself to citation, he would probably be thinking of work such as that of Bartlett, whose *Remembering* (1932) summarizes 15 years of previous research and publication on the relationship between perception and recall.

face-to-face interaction rests on a three-turn structure. I say something; you respond to it; and I can then use the third turn to decide whether your response is adequate and adequately connected to what I said first. In that turn, I can either decide to move on or rework (repair) my first turn and hope you will respond more satisfactorily or ask you to explain (account for) your failure to link your turn in second position to my first utterance. This is a dynamic structure: the second turn for me is the first turn for you so that you can examine what I do in the third turn, from my position, as a second turn for you — and then use your next turn as a third position to comment on what I have done:

My sequence		*Your sequence*
1. My statement	=	1. Your statement
2. Your response	=	2. My response
3. My review	=	3. Your review

In reality, the process is often somewhat more complicated. Other parts can be inserted in the sequence, and the third turn may be left empty if the speaker does not choose to use it. Nevertheless, this will make the basic point — that the meaning of my actions is not determined by me as their author but by the response of others and our subsequent negotiation.

This analysis is at the heart of the social scientists' difficulties with the idea of biological accounts of human behavior. The idea of a gene for violence presupposes that we know what violence is. Violence is actually a label that observers apply to behavior as the outcome of their application of a set of ideas current in a culture, and which they then respond to on the basis of that culture's notions about what to do about violent acts and how those notions might assemble into some idea of a 'violent person.' The argument has been pursued more fully in the context of addiction and drunkenness by Lindesmith (1947) and by MacAndrew and Edgerton (1970). Lindesmith (1947) pointed out that opiate addiction required that a person recognize the connection between the withdrawal or unavailability of the substance and his or her negative physical sensations. Where opiates were administered for straightforward pain relief, under the conditions of his time, that connection was not made and one could not say that addiction had resulted. Clearly, a physiologist might identify modifications to that person's biological processes, which could lead to an investigation of the disruptive impact of the substance on the person's normal functioning. However, there was no simple equivalence between those disruptions and the behavioral consequences of being recognized, by self or others, as addicted.

MacAndrew and Edgerton (1970) looked at the introduction of alcohol to the indigenous peoples of North America. They show that initially their response was one of puzzlement and a degree of disorientation. Alcohol consumption did not result in the acts that might have been expected if there were a simple relationship between physiology and behavior. These acts appeared after a period of time when the Native Americans had been able to observe European behavior under intoxication and to formulate their experiences in a comparable way with comparable behavioral consequences. Becker (1953, 1967) used a similar approach to discuss responses to cannabis and to LSD and the difference between the social experience of drugs knowingly consumed in a group environment and those that could be consumed unknowingly or in isolation. The ingestion of pharmacologically comparable substances does not lead to consistent and uniform behavioral effects in the way that biological determinism requires. The naïve user of drugs comes to learn what the experiences mean and how to act on the basis of them as a result of interaction with the sources of information and symbolic encoding available to them. These arguments can be extended further to consider normal body experiences more generally. We learn how to be well and how to be sick (Dingwall, 1976).

It is important, however, not to overstate this case. One of Mead's important contributions is his insistence on the materiality of embodiment. All social interrelations and interactions are rooted in a certain common socio-physiological endowment of every individual involved in them (Mead, 1962). As Dingwall (1976) stresses, the ability to operate 'normally' as a member of a particular socio-cultural group depends upon the consistency of one's physical endowments and functioning with the requirements of membership. This is an important distinction from more recent constructionist arguments. The material world is not explained away or treated as indefinitely pliable. Even a postmodernist cannot play soccer with a broken leg.

HUMAN BIOLOGY AND THE SOCIAL SCIENCES

What would it take, then, to produce a reconciliation between biologists interested in contributing to the explanation of human behavior and the social scientists who have regarded themselves as the experts in this field? Four elements seem likely to be involved in this:

First, there needs to be a more serious degree of mutual respect. As this paper has shown, this is a longstanding debate about serious issues in the interpretation of data and the philosophy of knowledge. It is not sufficient for biologists to dismiss the social scientific objections as the result of current posturing with wacky European theorists that have led to a normative disdain for the possibility of a biological contribution to the understanding of human societies. While there is more truth to this than we would like to think (or admit), we have shown in this paper that a long and learned history of engagement by serious scholars has been somewhat neglected. If social scientists have ceased to pursue their side of this debate, it may well have something to do with the biologists' failure to come up with better answers than they had in the 1920s. Should this be the case, then the social scientists' switch of attention to considering what social and material interests might be involved in propping up an intellectually questionable line of work may be forgivable, if not entirely justifiable. However, it is also patently the case that social scientists have abandoned any attempt to sustain this line of argument. From one year to the next a typical undergraduate would never encounter most of the authors we have cited or the arguments that they set out. It is notable that most of the debates about biological determinism have occurred between biologists and that other (social and natural) scientists have made very little contribution. Social scientists have a responsibility to consider whether there is anything in the new genetic understandings of human biology that requires a different response or, at the very least, a more informed restatement of the classic problems.

Second, biologists, especially those working at the molecular level, need a better understanding of environments. The sorts of generalizations that are being made on the basis of biological work are species-wide but are rarely subjected to adequate testing in relation to the diversity of environments under which our species is capable of flourishing. Those environments are not simply material, in the sense that an ecologist might recognize, but are also cultural and symbolic. A generalization about the relationship between genetic polymorphisms and behavior needs to be sustainable across the varying cultural frameworks that contribute to the understanding of that behavior. Murder is not the same as ritually prescribed killing or an unknowing act of witchcraft, although all may involve a violent death.

Third, and following from the above, biologists need a greater degree of specificity in the linkage between biology and behavior. The contrast between the precision with which genotypes and their physiological consequences are described and the looseness with which the social consequences are matched to them is striking. We have noted one example in the elision between "unwanted touching of genitals" and "touching of genitals" in Caspi et al. (2002), but this is not unique. More fundamentally, that paper makes a leap of inference from a set of indicators that suggest some people who, in an Australasian context, are more troublesome to be around than others to claiming a connection to violence. The latter, however, is a definition founded in local culture and applied in particular contexts by particular observers. It is a property ascribed to the behavior rather than inherent in it. We might also have explored the implication of Faris' (1921) arguments against just-so stories for much writing in evolutionary psychology.

Finally, it is equally important for social scientists to take biology more seriously. The dismissive fashion in which it has been treated since World War II does not do justice to the scale and subtlety of the body of work involved. We think, in particular, that Faris' (1921) comments on temperament are worthy of further reflection. It should not be a great point of contest for social scientists to accept that people's biological constitutions differ in ways that may have relevance to the material conditions for social interactions. If the Caspi et al. (2002) study (and studies like it) were read as a study of genetic polymorphism and temperament, it might be rather more persuasive. People may well have their neurological processes constructed and organized in ways that influence the speed with which they operate, their internal regulation, and the retention and recall of information — to name but three possibilities. Put into social situations, any of these may have an impact on process and outcome, although, as we have argued, that is unlikely to be a simple linear effect. It will, at the very least, be affected by the biological material that they encounter in the form of other people and their temperaments and by the symbolic resources shared by the participants (which provide the raw material out of which each interprets and responds to the others' behavior). The modest study of temperament may, however, be much less exciting and fundable than the alluring prospect of a pharmacological solution for criminal tendencies.

ACKNOWLEDGMENTS

We are grateful to Paul Martin and Anne Murcott for their helpful and constructive comments on earlier versions of this paper.

REFERENCES

Abbott, A., *Chaos of Disciplines*, University of Chicago Press, Chicago, 2001.

Ariés, P., *Centuries of Childhood*, Harmondsworth, Penguin, 1973.

Bartlett, F., *Remembering: A study in Experimental and Social Psychology*, Cambridge University Press, Cambridge, 1932.

Becker, H.S., Becoming a marihuana user, *American Journal of Sociology*, 59, 235–142, 1953.

Becker, H.S., History, culture and subjective experience: an exploration of the social bases of drug-induced experiences, *Journal of Health and Social Behavior*, 8, 163–176, 1967.

Blumer, H., *Symbolic Interactionism: Perspective and Method*, University of California Press, Berkeley, CA, 1969.

Bohannan, P., Ed., *African Homicide and Suicide*, Oxford University Press, Oxford, 1960.

Brunner, H.G., Nelen, M.R., Breakefield, X.O., et al., Abnormal behavior associated with a point mutation in the structural gene for monoamine oxidase A, *American Scientist*, 84, 132–145, 1993.

Caspi, A., McClay, J., Moffitt, T.E., et al., Role of genotype in the cycle of violence in maltreated children, *Science*, 297, 851–854, 2002.

Conrad, K., *Der Konstitutionstypus*, Springer, Berlin, 1963.

Darwin, C., *The Descent of Man and Selection in Relation to Sex*, Princeton University Press, Princeton, NJ, 1981. [reprint of 1871 first edition]

Darwin, C., *The Expression of The Emotions in Man and Animals*, John Murray, London, 1872.

Dingwall, R., *Aspects of Illness*, Martin Robertson, London, 1976.

Dingwall, R., Some problems about predicting child abuse and neglect, in *Child Abuse: Public Policy and Professional Practice*, Stevenson, O., Ed., Harvester Wheatsheaf, Hemel Hempstead, 1989, pp. 28–53.

Faris, E., Are instincts data or hypotheses? *American Journal of Sociology*, 27, 184–196, 1921.

Glueck, S. and Glueck, E., *Unraveling Juvenile Delinquency*, Harper and Row, New York, 1950.

Glueck, S. and Glueck, E., *Physique and Delinquency*, Harper and Row, New York, 1956.

Gould, S.J., *Ontogeny and Phylogeny*, Cambridge, MA: Harvard University Press, Cambridge, MA, 1977.

Graham, P., Dingwall, R., and Wolkind, S., Research issues in child abuse, *Social Science and Medicine*, 21, 1217–1228, 1985.

Hunter, H., YY chromosomes and Klinefelter's Syndrome, *Lancet*, 1(7444), 984, 1966.

Kay, L.E., *The Molecular Vision of Life*, Oxford University Press, New York, 1993.

Kretschmer, E., *Korperbau und Charakter*, Springer, Berlin, 1921. [translated as *Physique and Character: An Investigation of the Nature of Constitution and of the Theory of Temperament*, Kegan Paul, Trench and Trubner, London, 1925]

Lindesmith, A., *Opiate Addiction*, Principia Press, Bloomington, IN, 1947.

Lombroso, C., Introduction, in *Criminal Man According to the Classification of Cesare Lombroso*, Ferrara, G.L., Ed., Putnam, New York, 1911, pp. i–xv.

Lombroso, C. and Ferrero, W., *The Female Offender*, T Fisher Unwin, London, 1895.

MacAndrew, C. and Edgerton, R.B., *Drunken Comportment: A Social Explanation*, Nelson, London, 1970.

McDougall, W., *An Introduction to Social Psychology*, Methuen, London, 1909.

Mead, G.H., *Mind, Self and Society from the Standpoint of a Social Behaviorist*, University of Chicago Press, Chicago, 1962.

Mead, G.H., *Essays in Social Psychology*, Transaction Publishers, New Brunswick, NJ, 2001.

Plomin, R., The genetic basis of complex human behaviors, *Science*, 264, 1733–1739, 1994.

Rowe, D.C., *Biology and Crime*, Roxbury, Los Angeles, 2002.

Sheldon, W., *Varieties of Human Physique*, Harper and Row, New York, 1940.

Strauss, A. L., *Mirrors and Masks: The Search for Identity*, Martin Robertson, London, 1977. [first published 1959]

Taylor, I., Walton, P., and Young, J., *The New Criminology*, Routledge and Kegan Paul, London, 1973.

Thomas, W.I., The scope and method of folk-psychology, *American Journal of Sociology*, 1, 434–445, 1896.

Thomas, W.I., The gaming instinct, *American Journal of Sociology*, 6, 750–763, 1901.
Watson, J.B., *Behaviorism*. London: Kegan Paul, Trench, Trubner, London 1925.
Wittgenstein, L., *Philosophical Investigations*, Blackwell, Oxford, 1972. [first published 1953]

3

THE SCIENTIFIC AND CULTURAL MEANING OF THE ODIOUS APE–HUMAN COMPARISON

Jonathan Marks

CONTENTS

INTRODUCTION

The American lawyer Clarence Darrow — best known as the defender of John T. Scopes, prosecuted for teaching evolution in Tennessee in 1925 when it was newly illegal — lost the celebrated case, but successfully demonstrated what fools the opponents of biology were. Yet Darrow underwent an interesting transformation himself. Within a year he was publishing strident attacks on contemporary biology, in particular against genetics (Darrow, 1925, 1926), and had evolved from American biology's greatest public defender to its greatest basher. What transpired to bring this transformation about? The answer is probably very simple: Darrow actually read the textbook from which John T. Scopes was accused of teaching evolution: a book called *A Civic Biology* by George W. Hunter (1914).

Hunter's textbook was standard fare for the age. So standard, in fact, that shortly after its discussion of Darwinism and evolution, it moved on to the supremacy of the white race and more significantly to the necessity of sequestering and curtailing the reproduction of the antisocial and the "feebleminded." This culture, and which they placed Darrow, the great modernist, atheist, and freethinking civil libertarian, in a bind: he had just defended the teaching of science out of a textbook that advocated the curtailment of the rights, and indeed the very humanity, of the poor and defenseless whom he had devoted his professional life to defending. Tellingly, the very next case he would argue was that of a black doctor, Ossian Sweet, accused of murdering a member of the white rock-throwing mob outside his home in Detroit.

Darrow recognized that in spite of the obvious merits of science and of evolution, something was rotten in the state of Darwinism. He began to write articles criticizing the core of Darwinian eugenics, before any biologist (Pearl, 1927; Hogben, 1932) had either the nerve or the insight to do so. While the insight may seem an obvious necessity, the fact that it required nerve as well is attested by the fact that Raymond Pearl ended up having his offer of a professorship at Harvard retracted as a result of his public critique of eugenics (Glass, 1986).

The initial salvos against eugenics, consequently, were fired by non-biologists. "Amongst the schemes for remoulding society this is the most senseless and impudent that has ever been put forward by irresponsible fanatics to plague a long-suffering race," Darrow wrote (1926:137), never having been known as one to understate a case — and now the ideas, indeed the words, in the textbook at the very heart of the Scopes trial suddenly are those of "irresponsible fanatics."

Time indeed bore out Darrow's intuitions, but it is specifically Darrow's bind at the conclusion of the Scopes trial that interests me, because it is something that resonates with me as a student of human diversity and evolution and as a citizen of the modern world. On the one hand, the theory of evolution is good, or at least benign, and on the other, there are also sloppy theories of evolution that have odious implications, yet partake of the mainstream, and are for the most part tolerated or even accepted by biologists (e.g., Fisher, 1992; Buss, 1994; Wright, 1994; Thornhill and Palmer, 2000). Sociobiology and evolutionary psychology have today, like eugenics in Clarence Darrow's time, become Darwin's ventriloquists, if you believe their literature, and to challenge them, or even to challenge their wilder claims, is to invite the accusation of being a creationist (Alcock, 2001), no matter how much more you may oppose creationism.

The argument in the modern literature that concerns me here is a familiar one: Some people are like apes — not you or me, of course, but someone else. This argument is interesting anthropologically for the following reason. The illustration shown in Figure 3.1 is from 1824, a treatise on the human species by Julien-Joseph Virey. It may well be another predictable (and therefore somewhat passé) racist image, yet, although racist, it is far from passé, because it dates from a quarter century *before* Darwin's *Origin of Species* was published. From the standpoint of what has been called the anthropology of reason (Rabinow, 1996), we might therefore ask, "What can such an image mean outside the framework of a theory of evolution?" If no common descent or biohistorical relationship is implied, then what is the scientific meaning of this illustration? A reasonable answer might be found in *The Great Chain of Being* (Lovejoy, 1936). Obviously this is some sort of representation of the idea of the African as a connecting link between civilization and brute nature. The only problem is that, by 1824, within the scientific community, the *Great Chain* had lost its authority, owing principally to the work of Cuvier, following Linnaeus. Cuvier argued with great effect that there are *four* kinds of animals, they are all different, and are so different as to be noncomparable, much less linearly rankable (Appel, 1987).

Cuvier is enigmatic in that he reverses Linnaeus's decision and separates the human species taxonomically from all other primates, as the "Bimana" or two-handed ones; yet he readily likens Sarah Baartman, the so-called Hottentot Venus, to an ape (Schiebinger, 1993). The question I am posing is, "What can such a comparison mean scientifically to someone who believes neither in a theory of evolution nor in a *Great Chain of Being*?" The answer, I think, is that the comparison has *symbolic* force even to biologists who could not, and would not, maintain that it has *scientific* force. Such symbolic force seemed to be afforded by the work of anatomist Petrus Camper, whose adoption of a standardized approach to drawing faces (the facial angle) presented a gradation of form linking the ape to the European, through the African:

> When in addition to the skull of a negro, I procured one of a Calmuck, and had placed that of an ape contiguous to them both, I observed that a line, drawn along the forehead and upper lip, indicated this difference in national physiognomy; and also pointed out the degree of similarity between a negroe and the ape . . .. When I made these lines to incline forwards I obtained the face of an antique [European]; backwards, of a negroe; still more backwards, the lines which mark an ape, a dog, a snipe, &c. - This discovery formed the basis of my edifice. (Camper, 1794:9)

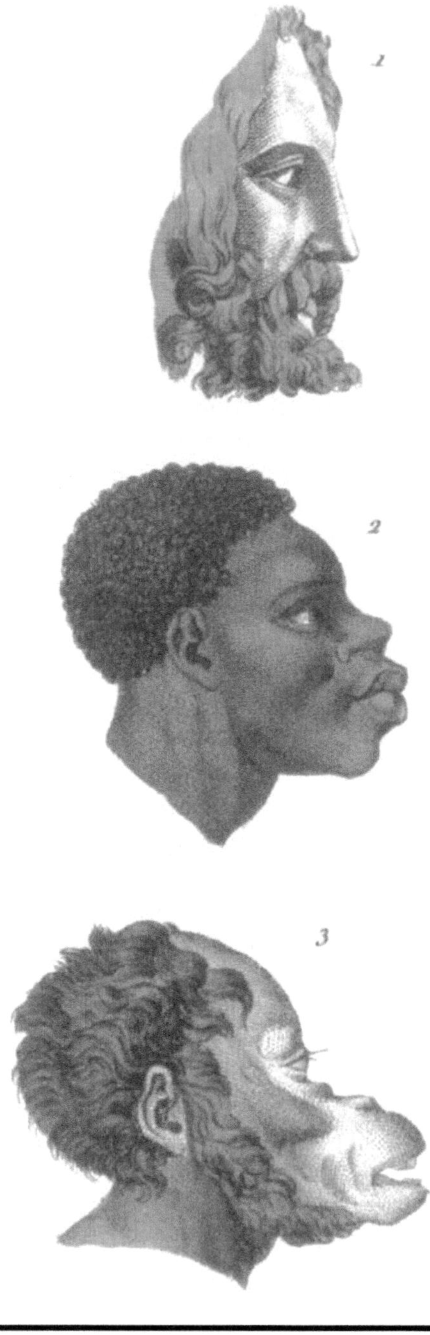

Figure 3.1 The intermediacy of the African between the European and the ape is implied in this illustration from the work of Julien-Joseph Virey in 1824, decades before the theory of evolution.

It is an irony of history that such was not his intent, for he strove merely to correct the artistic representations of non-Europeans, not to establish their place in the natural order (Gould, 1987). His work became widely known in the 1770s, and his illustration certainly seemed to suggest the physical intermediacy of Africans:

> The assemblage of craniums, and profiles of two apes, a Negro and a Calmuck, in the first plate, may perhaps excite surprise. The striking resemblance between the race of Monkies and of Blacks, particularly upon a superficial view, has induced some philosophers to conjecture that the race of blacks has originated from the commerce of the whites with ourangs and pongos; or that these monsters, by gradual improvements, finally become men.
>
> This is not the place to attempt a full confutation of so extravagant a notion. I must refer the reader to a physiological dissertation concerning the ourang-outang, published in the year 1782. I shall simply observe at present, that the whole generation of apes, from the largest to the smallest, are quadrupeds, not formed to walk erect; and that from the very construction of the larynx, they are incapable of speech. Further: They have a great similarity with the canine species, particularly respecting the organs of generation. The diversities observable in these parts, seem to mark the boundaries which the Creator has placed between the various classes of animals. (Camper 1794:32)

That Camper's "straw man" interpretations, which presupposed the instability and/or transmutation of species, were labeled explicitly as "superficial" and followed with an extended repudiation is less significant than that they were taken at face value and consistently presented within the scientific community. Camper himself inadvertently aided the process:

> It is amusing to contemplate an arrangement of these [skulls], placed in a regular succession: apes, orangs, negroes, the skull of an Hottentot, Madagascar, Celebese, Chinese, Moguller, Calmuck, and divers Europeans. It was in this manner that I arranged them in a shelf in my cabinet, in order that those differences might become the more obvious . . .." (Camper, 1794:50)

In Figure 3.2 the American polygenist, proslavery, creationist writers Josiah Nott and George Gliddon make the same point as Virey decades earlier.

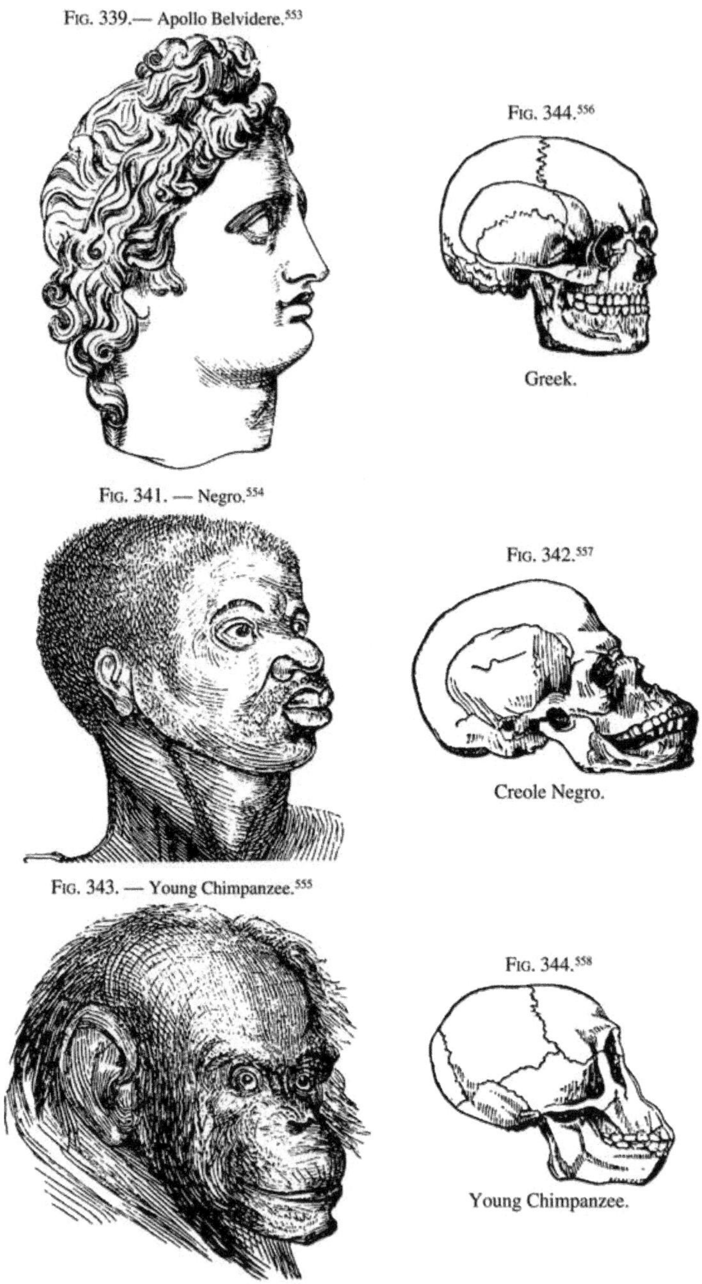

Figure 3.2 Josiah Nott and George Gliddon's 1857 *Indigenous Races of the Earth* reiterated the physical intermediacy of the African skull in espousing a proslavery, polygenist, creationist theory of human origins.

From the other side of the political spectrum, the relation of apes and human races remained largely intact. In *Vestiges of the Natural History of Creation*, Robert Chambers anonymously adopted a Lamarckian evolutionary theory that incorporated the relations within the human species itself. He wrote,

> All of these phenomena [of the diversity of human form] appear, in a word, to be explicable on the ground of *development*. We have already seen that various leading animal forms represent stages in the embryotic [sic] progress of the highest — the human being. Our brain goes through the various stages of a fish's, a reptile's, and a mammifer's brain, and finally becomes human. There is more than this for, after completing the animal transformations, it passes through the characters in which it appears, in the Negro, Malay, American, and Mongolian nations, and finally is Caucasian.
>
> *The leading characters, in short, of the various races of mankind, are simply representations of particular stages in the development of the highest or Caucasian type.* The Negro exhibits permanently the imperfect brain, projecting lower jaw, and slender bent limbs, of the Caucasian child, some considerable time before the period of its birth. The aboriginal American represents the same child nearer birth. The Mongolian is an arrested infant newly born. (Chambers, 1844:306–307, emphasis in original)

Chambers' work, while widely discussed, carried little weight in the scholarly community (Millhauser, 1959). Yet however ambivalent Charles Lyell may have been about evolution, he could write in *The Antiquity of Man* (1873), "The average Negro skull differs from that of the European . . . in all which points an approach is made to the simian type." Likewise, Thomas Huxley — abolitionist, monogenist, and now an evolutionist — explained the position of black people in the natural order in an 1865 essay:

> It may be quite true that some negroes are better than some white men; but no rational man, cognisant of the facts, believes that the average negro is the equal, still less the superior, of the average white man. And, if this be true, it is simply incredible that, when all his disabilities are removed, and our prognathous relative has a fair field and no favour, as well as no oppressor, he will be able to compete successfully with his bigger-brained and smaller-jawed rival, in a contest which is to be carried on by thoughts and not by bites. The highest places in the hierarchy of civilisation will assuredly *not* be within the reach of our dusky

> cousins, though it is by no means necessary that they should be restricted to the lowest. (Huxley, 1865:17–18; emphasis in original)

In other words, the position of black people, subordinate to whites — more apelike, more bestial, intermediate between European and ape — was largely unaffected by the Darwinian revolution, which readily appropriated the familiar imagery of pre-Darwinian racial thought. The nonwhite races embodied the connection between Europeans and the apes. The invocation of primates as a tool for dehumanizing outsiders — most infamously, the Irish and Africans — was simply transferred to an evolutionary discourse and thereby given exaggerated validity as science; preevolutionary images associating apes and "lesser races" were now simply vested with the authority of Darwinism.

Consequently, Figure 3.3 is a post-Darwinian image ostensibly showing the continuity of form between a "savage Hottentot" and a gorilla, but neither the southern African Khoe nor the ape is accurately rendered. Liberties are taken with both images to make the symbolic point that the African lacks humanity (Merians, 1994). Of course, in practice someone from *any* human population might well stand in for the imaginary Khoe and make the point at least as well, and perhaps better.

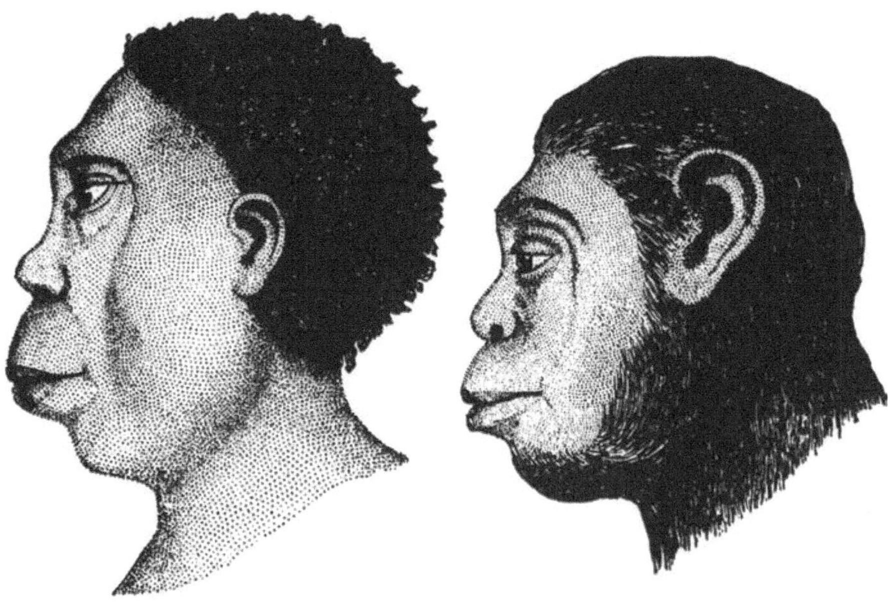

Figure 3.3 Post-Darwinian images continued to liken African people to apes, although taking liberties with the faces of both groups to make their point, as in this late nineteenth century illustration (unknown origin; collection of the author).

The intermediacy of the African between the European and the ape is probably best understood as an expression of Foucault's "bio-power," in which pervasive cultural forces have subtly naturalized the relationships between the races and, in conjunction with the anatomy of the ape, have established one race as superior and the others inferior (as they had long believed — a belief now legitimized by science and thus, in a sense, controlled). The human–ape juxtaposition then, independent of and unaffected by Darwinism, is a rhetorical tool for the symbolic dehumanization of unfavored people. It can also be as benign as a personal or political insult (see Jolly, 1972)* but it is certainly never a compliment and is loosely related at best to scientific inference.

DEMONISM IN THE EYE OF THE BEHOLDER

The work that inspired this essay was a popular book called *Demonic Males: Apes and the Origins of Human Violence*, coauthored by a distinguished Harvard primatologist (Wrangham and Peterson, 1996). The text is about war, male bonding, and general boyish mayhem being homologous in humans and chimpanzees. But when I was asked to review it I was struck most by the cover art, which reproduced a famous drawing by the Swiss anatomist Adolph Schultz (1933), illustrating the naked male bodies of a human, gorilla, chimpanzee, and orangutan, in this case cropped to emphasize the human and gorilla (Figure 3.4). The human in this case was the corpse of a 40-year-old black man Schultz had dissected at Johns Hopkins. Schultz's artwork had come under criticism in 1965 in a paper in the *American Anthropologist* for its penchant for comparing apes and blacks (Gould, 1965), thereby evoking the images we saw earlier. Physical anthropologists have commonly reproduced it in silhouette form, or at least downplayed the human image. When Ramona and Desmond Morris used the figure in their 1966 book *Men and Apes*, it was notably missing the "men."

Thirty years later, cropped to highlight the features of the man and gorilla, the illustration could only be considered provocative for its implications, particularly in light of the subtitle: *Apes and the Origins of Human Violence*. Schultz's anonymous black male cadaver from 1933 has come to symbolize not merely the demonic human male, but the very roots of human aggression made manifest. Happily the cover was rethought in the intervening few years, and now sports a tasteful abstract rendering.

It is also worth noting that Schultz's figure went through several published versions prior to this one. Initially, in 1926, Schultz published a figure comparing the body proportions of gibbon, orangutan, chimp, gorilla, and (a white) man; in 1932, he showed orangutan, chimp, gorilla, "Negro," and white (Figure 3.5), then finally eliminated the white for the

* Or the Web site www.bushorchimp.com

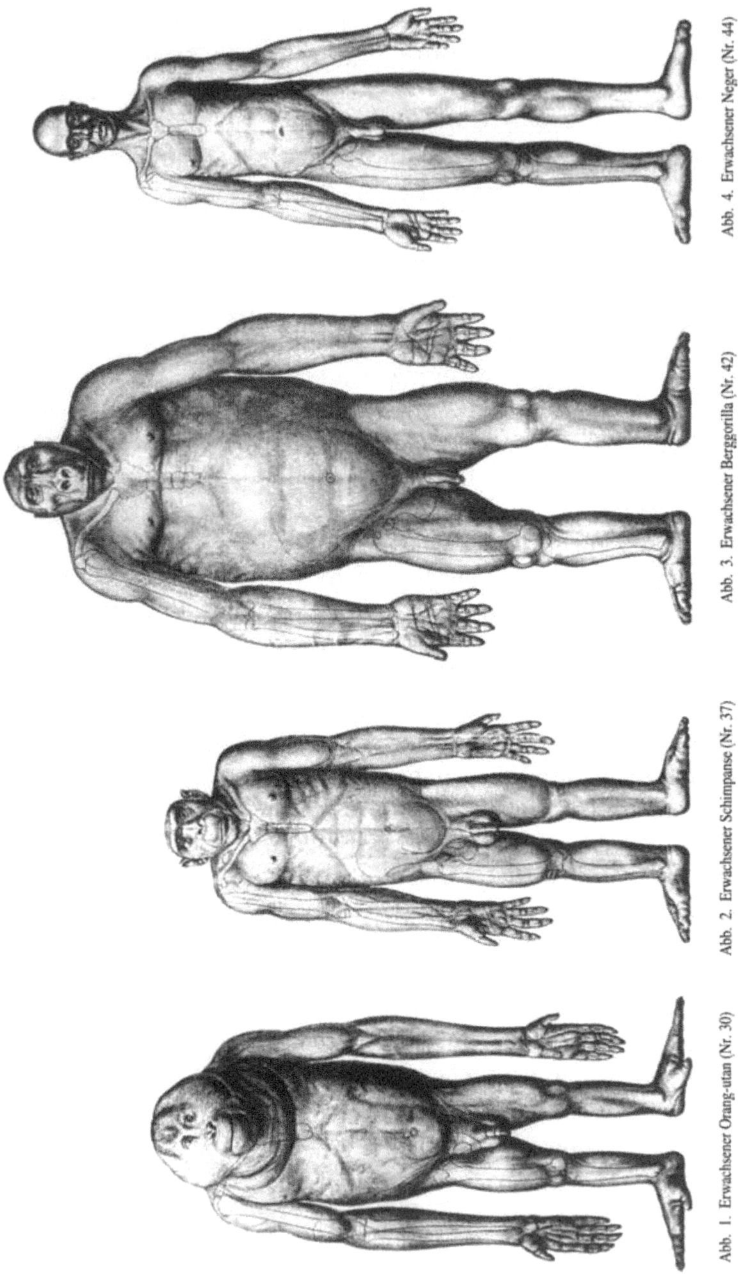

Figure 3.4 An illustration by the anatomist Adolph Schultz (1933) comparing the naked male bodies of a orangutan, chimpanzee, gorilla, and human (the corpse of a 40-year-old black man Schultz had dissected at Johns Hopkins) cropped to emphasize the human and gorilla.

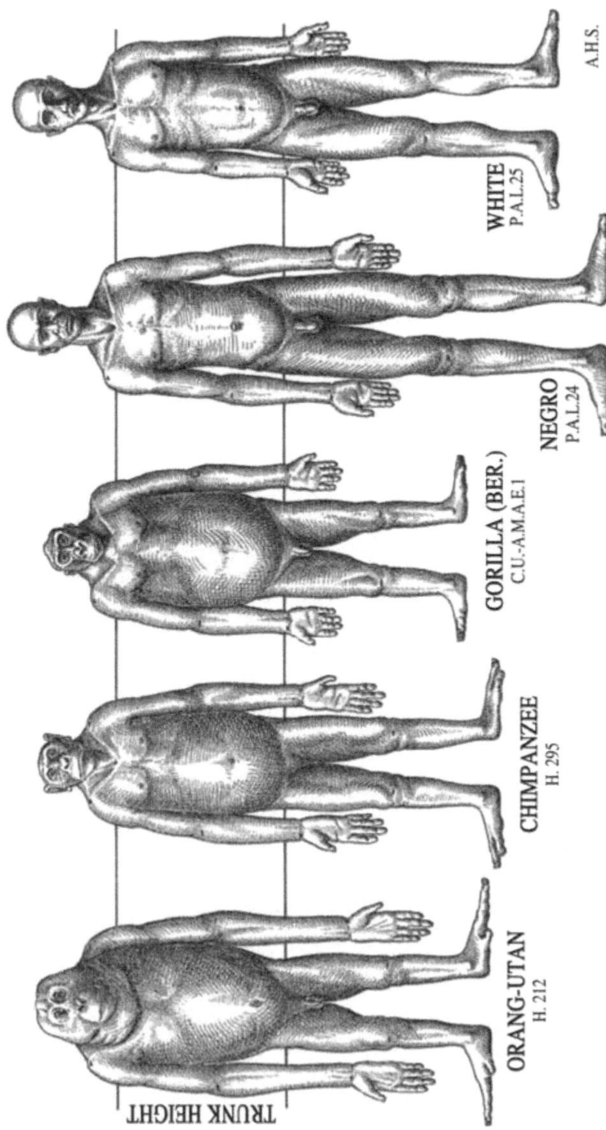

Figure 3.5. Adolph Schultz's earlier (1932) illustration comparing the body proportions of an orangutan, chimp, gorilla, "Negro," and white man (from which the white man was eliminated for the 1933 illustration; see Figure 3.4).

1933 illustration. It is certainly possible that the comparison of apes, with the black man representing humans, may ultimately have had the opposite meaning for Schultz as it did for other viewers. Decades later he claimed that he had intended his human figure to demonstrate the most extreme deviation from the ape-like condition (notably, long legs and narrow thorax; Schultz, 1966). This is the version he intended for posterity and included in his 1969 book, *The Life of Primates.* However, it would paradoxically undermine his goal of trying to illustrate "average" representatives of the species for comparative purposes. The illustration also contravenes Aleš Hrdlika's notorious proposal, articulated in 1908, that the principal aim of physical anthropology was "the study of the normal white man living under average conditions . . .."

Schultz could be damned for comparing a white man to the apes in the first version of the drawing (and thereby implicitly disregarding the rest of our species), for comparing a black man and then a white man to the apes in the second version (and thereby invoking a form of the *Great Chain of Being*, with blacks connecting apes to whites), and for comparing a black man alone to the apes (and thereby implying that blacks afford a more appropriate comparison to apes than do whites). Perhaps the most fundamental problem lies simply in the conflation of hominoid macroevolutionary diversity with human microevolutionary diversity. It is a conflation that is almost inescapable and yet culturally very powerful.

Indeed, along with the first incarnation of the illustration (which only illustrated the white man alongside the apes), Schultz (1926) included a list of ostensibly persistent features (from fetal life through adulthood) by which blacks and whites differed and by which one of them was more similar to apes. Blacks were more similar to apes in nine such features, and whites in only five. It is consequently difficult to escape the inference that in a universe of essentialized human racial forms, in 1926 the black man formed a connecting link between the white man and the ape.

EPISTEMOLOGY AND THE COMPARISON

It is worthwhile to explore the scientific inferences that can be drawn from finding human and ape to be similar. In the first place, you expect them to be similar, for they share an intimate biological history; if you find human and cow to be more similar in a particular sense than human and chimpanzee, that would be noteworthy. However, do behavioral commonalities between human and ape necessarily represent the same thing?

For example, let us consider infanticide. Some chimpanzees have sometimes done it; some humans have sometimes done it. Is it the same thing in both species? In both species you end up with dead babies. In chimpanzees it occurs generally in an aggressive context by nonrelatives, usually males, and culminates in an act of cannibalism. In humans it is generally perpetrated or sanctioned by the mother herself for economic or social reasons (Hausfater and Hardy, 1984). It certainly can be articulated as the same *word* in both species, but do they represent biological homologies or phonological homonyms? In other words, are they the same feature, inherited intact from a common ancestor or different features that just share a name? The answer is far from clear. Homology is a crucially contested concept in evolutionary biology and may be far more difficult to establish than simply observing superficially similar traits in related species (de Beer, 1971; Cartmill, 1994). The point is that it is by no means self-evident, nor is there a simple test one can conduct to decide. We have two behaviors in two closely related species, significantly similar yet meaningfully different.

This leads to an important conclusion: *Any human feature shared with an ape may not necessarily be illuminated at all by studying its manifestation in the ape.* Without a reasonable presumption that the feature is homologous — that is to say, the product of passive heredity in both lineages from an ancestral form — it can only be meaningful as metaphor.

Demonic Males, for example, tells us with extraordinary biological opacity, "We could well substitute for Sparta and Athens the names of two male chimpanzees . . .." (Wrangham and Peterson, 1996:192). Yes, of course we *could*, but since Boo-Boo and Bam-Bam have no particular biological connection to ancient Hellenic political entities, it follows that we learn nothing greater scientifically about the Peloponnesian War by substituting the names of two male chimpanzees than we would learn by substituting the names of two soft drinks, two subatomic particles, or two immiscible liquids of different specific gravities. There is, indeed, some truth in the recognition that when entities interact, one may ultimately predominate or prevail in some fashion, but that wisdom is hardly biological. The point is that what looks meaningful scientifically in this context, in fact, derives its meaning metaphorically. This relationship is literary, evocative, and rhetorical — and, thus, quintessentially a product of human evolution — but it is not especially scientific and is negligibly Darwinian.

Along similar lines, we find the Great Ape Project, which brilliantly unified animal rights activists and primate conservationists, to end up promoting the idea of human rights for the great apes (Cavalieri and Singer, 1993) — in particular, the rights to life, liberty, and freedom from torture. Now of course no sane person advocates death, imprisonment, and torment for apes. I would not deny for an instant the necessity to defend primates in captivity and protect them in the wild. The question is, how does an animal rights issue become transformed into a human rights issue?

The most compelling and resonant argument is derived from the cognitive abilities of apes (paradoxically established while denying apes their liberty).

Here we learn that apes are explicitly like learning-disabled people: sometimes severely mentally handicapped, sometimes merely autistic, sometimes merely young. Yet the learning-disabled get human rights, so why not the apes? The problem with this can be readily seen (Groce and Marks, 2000): if you believe apes are like learning-disabled humans, then do you reciprocally believe that cognitively impaired people are like apes?

We need hardly mention the analogies to African slavery that are common in this often-well-meaning but problematic literature. Preevolutionary developmentalism often saw technological primitiveness and biological primitiveness as linked. This drew on medieval traditions of monkeys as fools and foolish people reciprocally as monkeys. In the nineteenth century, it yielded facile connections among savages, children, and nonhuman primates (Jahoda, 1999). Once again, let me make explicit that I am very much in favor of the proper treatment of captive apes and the preservation of their wild habitats. My point is simply that dehumanizing people, especially the most vulnerable and historically dehumanized ones, simply does not afford a compelling moral basis for attaining that end. More to the point, it is scientifically invalid, insofar as learning-disabled people are biological people, and super-intelligent apes are metaphoric people.

Neither of which is to deny the power or the value of metaphor in science. 'Natural selection' and the 'genetic code' are two of the most historically profitable such devices still in use. The point I wish to explore here is not science as rhetoric, but science as some sort of positive knowledge — as beneficial, or at least as benign, knowledge. This leaves us with a dilemma. Basic evolutionary theory dictates that all people are equidistant from the nearest nonhuman species and differ from them by virtue of being part of an evolutionary unit we call *Homo sapiens*. What then can we learn about humans from observing apes while concurrently retaining the essential humanity of all humans? How do we learn about human evolution from studying chimpanzees at all, with a maximum of scientific rigor and a minimum of metaphoric flourish and scientifically meaningless and offensive imagery?

APES AND HUMAN EVOLUTION

We observe chimpanzees and we observe humans. If we find them to match in the feature we are studying, then we might as well study the humans, who are ubiquitous and can at least give their consent. What would studying chimpanzees tell us in this case about humans that we do not already know or that we cannot learn more easily by studying humans?

If, however, we find human and chimpanzee to be different, which of course we usually do, we have three evolutionary possibilities. First, either the chimpanzee state is ancestral and the human state is derived, in which case we can learn something about human biohistory from the ape, because the ape state is the antecedent of the human state. Alternatively, perhaps the human state is ancestral and the chimpanzee state is derived, in which case we cannot learn about human biohistory from the ape, because the human state is the antecedent and the modern ape state is thus irrelevant to human biohistory. Third, they are both derived separately from an unknown ancestral form, in which case, again, we learn nothing about human biohistory from the living ape.

Only in the first case, where to study the ape is to study a feature as it existed in our own ancestor, do we gain knowledge of the origin of the human condition from the ape. Yet the ambiguity I have just described is the current state of knowledge of something well-studied and accessible — like locomotion (Richmond et al., 2001) — not something vague and fluffy like the occasional and wildly over-interpreted behavior of some people and some apes.

Quite possibly the most interesting comparisons to be made today are not between the chimpanzee and human but between the two species of chimpanzees, which seem to differ in odd ways, the bonobo being more sexual and less of a carnivore or tool-user, for instance. This contrast tells us about humans only indirectly, but it tells us something important, namely, that we cannot assume the generalized behavior of common chimpanzees to be an ancestral state when it differs markedly from that of its sister taxon.

I have not even mentioned the gorilla, which may be the sister taxon of the chimpanzees (Schwartz, 1984) or equidistant from the chimps and humans (Marks, 2002) or may be the sister taxon of a chimp–human clade (Wildman et al., 2003). Either way, the gorilla is sufficiently behaviorally different from all three species to make a hash of any evolutionary speculation on its relation to human behavior.

CONCLUSIONS

We cannot study human evolution without comparing ourselves to the apes, but the apes are themselves not merely biological entities but cultural constructions — meaning-laden and symbolically powerful. When they are compared specifically to the apes, particular humans are thereby diminished in their humanity, a fact recognized by racists and exploited by political satirists for centuries. Anthropologists must be sensitive to the cultural message of ape–human comparisons; human biologists must be prepared to evaluate the evolutionary competence of such work, for, as in the time of Clarence Darrow and the "Monkey Trial," evolutionary theory is a precious baby, but there comes a point at which we need to jettison the bathwater.

REFERENCES

Alcock, J., *The Triumph of Socio-biology*, Oxford University Press, New York, 2001.

Appel, T., *The Cuvier-Geoffroy Debate: French Biology in the Decades Before Darwin*, Oxford University Press, New York, 1987.

Buss, D., *The Evolution of Desire*, Basic Books, New York, 1994.

Camper, P., *The Works of the late Professor Camper on the Connexion between the Science of Anatomy and the Arts of Drawing, Painting, Statuary, etc.*, Cogan, T., trans., Dilly, London, 1794.

Cartmill, M., A critique of homology as a morphological concept, *American Journal of Physical Anthropology*, 94, 115–123, 1994.

Cavalieri, P. and Singer, P., Eds., *The Great Ape Project: Equality Beyond Humanity*, Fourth Estate, London, 1993.

Chambers, R., *Vestiges of the Natural History of Creation*, John Churchill, London, 1844.

Darrow, C., The Edwardses and the Jukeses, *The American Mercury*, 6, 147–157, 1925.

Darrow, C., The eugenics cult, *The American Mercury*, 8, 129–137, 1926.

de Beer, G., *Homology, an Unsolved Problem*, Oxford University Press, London, 1971.

Fisher, H.E., *Anatomy of Love: The Natural History of Monogamy, Adultery, and Divorce*, Norton, New York, 1992.

Glass, B., Geneticists embattled: their stand against rampant eugenics and racism in America during the 1920s and 1930s, *Proceedings of the American Philosophical Society*, 130, 130–154, 1986.

Gould, L., Negro = Man, *American Anthropologist*, 67, 1281–1282, 1965.

Gould, S.J., Petrus Camper's angle: the grandfather of scientific racism has gotten a bum rap, *Natural History*, 96, 12–16, 1987.

Groce, N.E. and Marks, J., The Great Ape Project and disability rights: ominous undercurrents of eugenics in action, *American Anthropologist*, 102, 818–822, 2000.

Hausfater, G. and Hardy, S.B., Eds., *Infanticide: Comparative and Evolutionary Perspectives*, Aldine, New York, 1984.

Hogben, L., *Genetic Principles in Medicine and Social Science*, Alfred A. Knopf, New York, 1932.

Hunter, G.W., *A Civic Biology, Presented in Problems*, American Book Company, New York, 1914.

Huxley, T.H., Emancipation: black and white, In *Man's Place in Nature, and Other Anthropological Essays*, Huxley, T.H., Ed., Macmillan, New York, 1865.

Jahoda, G., *Images of Savages: Ancient Roots of Modern Prejudice in Western Culture*, Routledge, New York, 1999.

Jolly, A., *The Evolution of Primate Behavior*, Macmillan, New York, 1972.

Kuhl, S., *The Nazi Connection*, Oxford University Press, New York, 1994.

Lovejoy, A.O., *The Great Chain of Being*, Harvard University Press, Cambridge, 1936.

Lyell, C., *The Antiquity of Man*, 4th ed., John Murray, London, 1873.

Marks, J., Genes, bodies, and species, In: *Physical Anthropology: Original Readings in Method and Practice*, Peregrine, P.N., Ember, C.R., and Ember, M., Eds., Prentice-Hall, Englewood Cliffs, 2002, pp. 14–28.

Merians, L.E., What they are, who we are: representations of the "Hottentot" in eighteenth-century Britain, *Eighteenth-Century Life*, 17, 14–39, 1994.

Millhauser, M., *Just Before Darwin: Robert Chambers and Vestiges*, Wesleyan University Press, Middletown, CT, 1959.

Morris, R. and Morris, D., *Men and Apes*, McGraw-Hill, New York, 1966.
Nott, J.C. and Gliddon, G.R., *Indigenous Races of the Earth*, J.B. Lippincott, Philadelphia, 1857.
Pearl, R., The biology of superiority, *The American Mercury*, 12, 257–266, 1927.
Rabinow, P., *Essays on the Anthropology of Reason*. Princeton University Press, Princeton, NJ, 1996.
Richmond, B.G., Begun, D., and Strait, D., Origin of human bipedalism: the knuckle-walking hypothesis revisited, *Yearbook of Physical Anthropology*, 44, 70–105, 2001.
Schiebinger, L., *Nature's Body*, Beacon, Boston, 1993.
Schultz, A.H., Fetal growth of man and other primates, *Quarterly Review of Biology*, 1, 465–521, 1926.
Schultz, A.H., Man as a primate, *Scientific Monthly*, 33, 385–412, 1931.
Schultz, A.H., Die Körperproportionen der erwachsenen catarrhinen Primaten, mit spezieller Berücksichtigung der Menschenaffen, *Anthropologischer Anzeiger*, 10, 154–185, 1933.
Schultz, A.H., A problem of labels, *American Anthropologist*, 68, 528, 1966.
Schwartz, J.H., The evolutionary relationships of man and orang-utans, Nature, 308, 501–505, 1984.
Thornhill, R. and Palmer, C., *A Natural History of Rape: Biological Bases of Sexual Coercion*, MIT Press, Cambridge, 2000.
Virey, J.J., *Histoire Naturelle du Genre Humain,* Crochard, Paris, 1824.
Wildman, D.E., Uddin, M., Liu, G., et al., Implications of natural selection in shaping 99.4% nonsynonymous DNA identity between humans and chimpanzees: enlarging genus *Homo*. *Proceedings of the National Academy of Sciences USA* 100, 7181–7188.
Wrangham, R. and Peterson, D., *Demonic Males: Apes and the Origins of Human Violence*, Houghton Mifflin, Boston, 1996.
Wright, R., *The Moral Animal: Evolutionary Psychology and Everyday Life*, Vintage Books, New York, 1994.

4

EVERYDAY EXPLANATIONS OF DIVERSITY AND DIFFERENCE: THE ROLE OF LAY ONTOLOGIZING

Kevin Durrheim and John Dixon

CONTENTS

In the streets, in cafes, offices, hospitals, laboratories, etc. people analyse, comment, concoct spontaneous, unofficial "philosophies" which have a decisive impact on their social relations, their choices, the way they bring up their children, plan ahead and

> so forth. Events, sciences and ideologies simply provide them with "food for thought." (Moscovici, 1984a:16)

INTRODUCTION

The development of the sciences during modernity changed common sense and everyday life fundamentally. Ever since science stopped being theological, suggests de Certeau (1984:6), "it constituted the whole as its remainder," creating common sense anew as a territory to be conquered. Of course, this organizing cleavage of modernity is tenuous. Sociologists of science and other commentators inform us that scientific projects are founded on everyday knowledge, which provides first-degree concepts and values and even cognitive strategies for second-degree scientific constructions (Flick, 1998; Schütz, 1966).

The sciences have also had a more direct influence on common sense as scientific concepts have trickled down and been incorporated into conversations and the thinking of ordinary people (Moscovici, 1984b). The person on the street today — and that includes all of us outside our small field of specialization — has opinions about a plethora of objects and processes that are the inventions of modern science. Ordinary conversations can turn to talk about the unconscious and neurosis, microwaves, neurotransmitters, black holes, the human immune system, genes, and genetic modification.

The aims of this chapter are twofold: first, to provide an overview of the way in which lay publics explain human diversity and difference and second, to show how such explanations provide the foundations for *racial projects*. This study of "racial projects connects what race *means* in a particular discursive practice to the ways in which both social structures and everyday experiences are *organized*, based upon that meaning" (Omi and Winant, 1994:56; emphasis in the original). Using interview material, we show how racial contact and segregation in postapartheid South Africa are organized around explanations of racial difference that are drawn from a variety of modern sciences. To locate the discussion we first present Moscovici's theory of modern common sense and then we review the small body of social psychological research that has investigated how ordinary people explain racial diversity.

THE PRODUCTION OF COMMONSENSE

Whereas theological society was governed by a distinction between the sacred and the profane, Moscovici (1984a) argues that modern society is governed by a distinction between the consensual and the reified universes. The consensual universe is that of everyday society, where individuals are free to compete with each other for meaning and where the implicit stock of images and ideas that makes up the content of thinking are produced and traded in conversation. In the reified universe, the various sciences impose their authority on meaning by deciding what is and is not true. Whereas the earlier religious ideology had sought to banish the profane, the work of ideology in the modern period is to "facilitate the transition from one world into the other" (Moscovici, 1984a:23). Science casts the consensual into reified categories, and by popularization, scientific categories enter into the consensual universe. This form of common sense, which has its roots in, and gains its authority from, science, Moscovici terms *social representation*.

On the basis of his study of the spread of psychoanalytic ideas, Moscovici suggests that this transition from the reified to the consensual universe, by which social representations are produced, takes place by means of two processes: anchoring and objectification. Anchoring is a mechanism whereby strange or new ideas — including, but not restricted to, scientific productions — are reduced to ordinary categories and images. The classifying and naming of new or unfamiliar objects integrates them into preexisting systems of thought and thus become part of common sense. Whereas anchoring is a universal process, objectification is particular to modern societies (Billig, 1988). Objectification also turns the unfamiliar into the familiar, but it does this by objectifying or materializing abstractions. The common sense of modern society is an objectified consciousness in which life and the world — formerly transcendentalized into religious or mythological abstractions — are made into concrete features of perceptible reality and the world of objects. Common sense today is "science made common" (Moscovici, 1984a:29).

The implications of this theory of common sense are profound, for objectification changes not only our cognitions, but also our world. As science produces knowledge and this trickles down into common sense, so the objectified world of reality is produced, and it is in this world that people must orientate themselves and act:

> The evidence of particular men has become the evidence of our own senses, an unknown universe is now familiar territory. The individual, in direct contact with this universe without the mediation of experts or their science, has progressed from a secondary to a primary relationship with the object, and this indirect assumption of power is a culturally fruitful action. (Moscovici, 1976:41; cited in 1984a)

Objectification is culturally fruitful, for it allows ordinary people in the consensual universe to speak in the name of reality, forgetting precisely that the object*ified constructs* of common sense are *social* creations. Social representations change our reality, our relations with reality (we can know it and speak the truth about it), and our relations with each other (we argue about reality in the name of truth).

The importance of this work on social representations lies not only in showing us that we have developed a uniquely modern way of seeing and thinking about the world, but that we have a new way of living and acting in it. Social representations provide a framework and reasons for action:

> Who does not have a representation which allows them to understand why liquids rise in a container, why sugar dissolves, why plants need to be watered, or why the government raises taxes? Thanks to this popular physics we avoid collisions on the road, thanks to this popular biology, we cultivate our garden, and this popular economics helps us look for a way of paying less tax. (Moscovici, 1998:236)

RACISM AND EXPLANATIONS OF DIFFERENCE

By and large, the social sciences have investigated people's descriptions rather than their explanations of human difference. Since abandoning the project of documenting racial differences (in intelligence in the 1920s), mainstream social psychology, for example, has been concerned with prejudiced attitudes, cognitive distortions, and racial stereotypes (Samelson, 1978). People's depictions of racial difference were treated as diagnostic indicators of cognitive biases and prejudiced minds. Given this focus, the extant body of literature on everyday explanations of human diversity and difference is small, comprising only a handful of studies. These will be reviewed here, for they provide two important lessons about everyday explanations of diversity and difference: (1) that ordinary people do have theories about the processes that produce and maintain human diversity; and (2) that these explanations provide both a rationale for, and a constraint to, action. Explanations of diversity and difference shape the kinds of racial projects that a society will undertake.

The impetus for studying explanations of diversity came from the puzzling picture of ambivalent racial attitudes that emerged in the United States by the late 1960s. On the one hand, political and legislative change in that country had produced what Pettigrew (1979) described as a "dramatic liberalization of racial attitudes." The fading of Jim Crow racism, however, had not translated into a change in behavior. American society remained discriminatory and segregated, and the same people who supported desegregation and equality as matters of principle opposed policies that aimed to make these ideas a reality (Bobo et al., 1997). If racial prejudice was a thing of the past, what could explain this persistent opposition to social transformation?

In their path-breaking work, Apostle et al. (1983) suggested that explanations of difference, not prejudiced racial beliefs, were critical:

> The ways people respond to an outgroup are less dependent on perceived differences than on how the perceived differences are explained. Thus, for example, if blacks are perceived as lazy, the attitude indicated by the perception is quite different if the laziness is accounted for as an innate black character trait than if it is understood to be the result of blacks having been denied for many generations access to opportunities to achieve. (Apostle et al., 1983:15–16).

Was it possible that the observed change in attitudes did not signal the elimination of racial prejudice but a change in explanations of difference? Earlier, Hyman and Sheatsley (1956) had suggested that the liberalization of attitudes in the United States may have been due to a "revolutionary change" that had taken place in beliefs about racial biology and intelligence, that is, in their explanations of difference: "Once the educability of the Negro has been granted, it becomes considerably more difficult to argue against integration in the schools" (Hyman and Sheatsley, 1956:19). During the 1980s a series of papers explored the possibility that explanations of difference could account for resistance to social transformation.

On the basis of in-depth interviews with 200 residents of San Francisco, Apostle et al. (1983) distinguished six "modes of explanation," each of which locate the cause of racial difference in a different ontology (see Table 4.1). This list of explanatory modes was similar to the set of "stratification beliefs" that had been included in national surveys since the late 1970s (Kluegel and Smith, 1982, 1983, 1986). In these studies, investigators wanted to determine how black inequality in terms of jobs, housing, education, and so on, was explained. These explanations of

Table 4.1 Explanations of Racial Difference

Mode of Explanation	*Causal Agent*
Supernatural	God
Genetic	Biology
Individualistic	Individual free will
Radical	White oppression
Environmental	Social structure
Cultural	Roots and culture

difference were categorized into two modes: individualistic modes that blame blacks themselves (e.g., innate ability or personality traits such as lack of motivation) and structuralist modes that blame the system (e.g., lack of opportunity for education and discrimination).

A substantial amount of empirical evidence has been garnered for the view that explanations shape racial policy opinions (Apostle et al., 1983; Bobo and Kluegel, 1993; Kluegel and Smith, 1982; Kluegel, 1985, 1990). For example, Apostle et al. (1983) report that individuals who believe in the radical or environmental mode support government intervention with policies such as affirmative action. However, individuals whom they call "Supernaturalists" or "Geneticists" tend to endorse expressions of traditional racial attitudes (i.e., support for laws against intermarriage and the right to segregate) and oppose affirmative action and government spending and assistance to help blacks.

This research has also made an important contribution to our understanding of the puzzling mismatch between the liberal attitudes of many white Americans and their opposition to social change. Kluegel (1990) argues that during the middle years of the twentieth century there was a significant decline in beliefs about the innate inferiority of blacks and hence a shift away from support, in-principle, for traditional blatant racism, such as *de jure* school segregation. However, these biological theories of difference were not replaced by structuralist explanations and support for remedial action. Rather, new individualistic explanations emerged that explained inequality in terms of psychological and motivational factors. Black disadvantage was attributed to laziness or some other "cultural pathology" (Goldberg, 1998), and thus the responsibility for change was placed on the shoulders of individual blacks and not policies such as affirmative action.

In sum, this small body of investigations into explanations of difference has shown, first, that ordinary people use distinct modes of explanation and, second, that explanations become socially efficacious because they provide a seemingly logical rationale for action.

LAY ONTOLOGIZING

Social psychological investigations of explanations of diversity offer some robust and theoretically fecund observations. The evidence confirms the hypothesis that explanations shape racial thinking and action. In this section we highlight some of the limitations of this traditional approach and propose a new framework for investigating everyday explanations of diversity and difference, drawing on the work of Moscovici and newer qualitative traditions of investigation in the social sciences.

The limitations of the small body of traditional research stem from certain assumptions about explanations that are presupposed or implicit in the quantitative methodology (Danziger, 1985). Generally the research proceeds as follows: participants are asked to indicate whether they believe (yes or no) in each of a list of explanations of racial difference or inequality; correlational methods are then used to determine whether individuals who endorse different explanations also have different beliefs about what can or should be done regarding social transformation. This has shaped the understanding of everyday explanations of diversity in two ways:

Unitary explanations: Each explanation is assumed to be coherent, and the different types of explanation are assumed to be distinct. The method of measurement *requires* that the explanatory modes be separated into distinct ideal types.

Individual differences: Apostle et al. (1983) refer to a typology of individuals (radicals, supernaturalists, individualists, etc.). Although authors have acknowledged that an individual may endorse more than one mode of explanation, such variability in explanation has not been the focus of investigation. On the contrary, individual differences are generally assumed by the correlational methodology employed. Individuals who endorse a particular combination of explanations are treated as a distinct type.

These assumptions may or may not be true. However, we need a new perspective since the quantitative correlational method forecloses the investigation by assuming the truth of unitary explanations and individual differences. We use Moscovici's account of social representations as the starting point for our investigation of everyday explanations because he recommends a qualitative methodology that is explicitly attuned to investigating the socially shared and interconnected nature of common sense. Moscovici (1973) argues that common sense consists of interconnected theories or cognitive systems in which ideas, beliefs, and explanations are interrelated with other ideas, beliefs, and explanations. These social representations originate and are reproduced in conversation and are thus shared by groups.

Moscovici recommends the use of historical and anthropological methods (Moscovici, 1988). Research should "concern itself with concrete analysis of theories, ways of thinking and speaking, conventional wisdom, all of which are distinctive features of the collectivity at a given point in time" (Muscovici, 1984b:948). Rather than approaching research deductively, providing respondents with belief statements to endorse, Moscovici recommends an inductive and descriptive approach. The aim of the analysis should be to document the theories (content) that people use to make sense of and construct their worlds and to investigate these qualitatively by studying how social representations are developed in conversations. Although Apostle et al. (1983) began their research with 200 interviews, they did not exploit this qualitative material to the full, but instead treated this work as preliminary to the real task of measurement. Their main aim was to "establish, in a more rigorous fashion . . . that the explanatory element is as central to the measurement of white racial attitudes as we claimed" (Apostle et al., 1983:36).

In our earlier work (Durrheim and Dixon, 2000), we recommended a focus on lay ontologies rather than social representations. Although the notion of social representations provides a useful corrective to the methodological and theoretical individualism of traditional social psychology, there are two reasons why it is not entirely adequate to the task of investigating explanations of difference:

First, Moscovici suggests that the rise of the sciences changed common sense fundamentally, and that it is no longer common sense, but social representation. In contrast, Billig (1988) argues that the contrast between social representations and common sense should be open to investigation to determine (1) whether objectified social representations appeared before the modern age and (2) whether nonobjectified or transcendental features of modern common sense exist. Billig suggests that it is premature to foreclose empirical investigation of the nature of common sense by restricting the focus of investigations to social representations. The term *lay ontologies* is suited to such a task. Like the notion of social representations, lay ontologies draws attention to specific ontological features of lay theorizing, thus reflecting the idea that common sense constructs reality, providing an account of what is. However, there is no reason why investigations of lay ontologies should be restricted to the objectifying ontologies characteristic of social representations. Studies of lay ontologies can thus facilitate the empirical task of determining whether the transformations from science have swept all before them or whether older symbolic systems remain (Billig, 1988).

The second restriction of the concept of social representations also marks the concept of lay ontologies — both concepts are nouns, deflecting attention away from the interactional features of conversations and the activity of construction. This is one of the major criticisms of investigations of social representations: "There has been almost no concern with what talk is doing and how it is arranged interactionally" (Potter and Wetherell, 1998:141). In addition to investigating the content of common sense, the newer perspectives of discourse analysis (Potter and Wetherell, 1987) and rhetoric (Billig, 1996) recommend a focus on everyday *language practices.* When people develop particular accounts of the world in their conversations, they are not simply describing the world but (1) actively constructing it, (2) in such a manner as to serve interactional ends in the conversation. Lay ontologies are used to answer questions, make accusations and justify beliefs and practices, and so on. Studies of rhetoric suggest that while common sense is shared, it is also divided or dilemmatic and that versions of the world are specifically designed to undermine alternative argumentative positions. Lay ontologies are developed and used in the context of justification and criticism and turn out to be very useful to these rhetorical purposes since they allow conversationalists to ground their arguments ontologically (i.e., in accounts of what is). Our investigation of common sense should thus be focussed on the conversational, rhetorical, and ideological activities of lay ontologizing.

We previously defined lay ontology as the "body of ontological 'discursive content' that is manifest as a 'chain of reasoning' about the nature of human nature, the world, or reality . . .. The concept of 'lay ontology' thus directs us empirically to a particular body of (ontological) discursive content, and to the reasoned elaboration of this content in theorizing about natural events, states, processes and outcomes" (Durrheim and Dixon, 2000:98). We want to add that in addition to tracking the unfolding content of ontological arguments across conversational utterances, we explore also the conversational, rhetorical, and ideological functions being achieved by the activity of lay ontologizing.

CHARACTERISTICS OF ORDINARY EXPLANATION OF DIFFERENCE

The remainder of the chapter will outline the central characteristics of the lay ontologizing used in everyday explanations of racial diversity. These accounts were developed in interviews with white South Africans about racial contact, desegregation, and transformation in that country. The full data consisted of 51 in-depth interviews conducted with individuals, groups, and families on Scottburgh beach during peak tourist season in the summer vacation of 1998–1999 and 2001–2002.* The beach was chosen as the interview site to ground the discussion of contact and experiences of desegregation in a geographical context that was historically segregated by apartheid legislation and that has conserved patterns of segregation and avoidance despite legislative and political change (Dixon and Durrheim, 2003; Durrheim and Dixon, 2001). The lay ontologizing reported in the extracts below was developed in accounts of why the interviewees

* A detailed account of the methods and ethics of this study can be found in Durrheim and Dixon (2005).

Figure 4.1 **Symbols of so-called 'petty apartheid': segregation of civic amenities by race (George Ellis, with permission).**

personally, and white South Africans generally, segregated themselves from other "race" groups.*

Multidimensional Causal Ontology

Extract 1

Jack: Their culture is a bit different to what (.) to what our culture is (Res - ja) I mean, how do you treat them? (0.5) It's like thi::s inheritance w::ell gene we a::ll (.) we all got (.) and you have to be scared when they're in masses hey (.) (Res - ja) But (.) here [on the Scottburgh beach] it's fine I mean (0.5) n::o problem (Res - mm) . . . It's like I said, it's this inherent gene we a::ll got from our forefathers that stayed here (Res - ja) when they're in masses hey (0.5) you gonna be scared obviously.

Extract 2

Joshua: I mean what would have frightened a hand full of/of Boers (Res: ja) more than: watching a whole impi of/of flipping *swart gevaar* [black peril] coming (Res: ja/ja) coming towards you (Res: ja/ja) Hey it must be scary hey (Res: ja) Let's face it (Res: ja/ja) a::nd lets also face it they were brave enough to stand up to them (Res: ja) (it gave a different)/the fact that they:: had rifles and um .hh the/the impi's had (.) um:::=

Res:=Assegais

Joshua: Assegais .hh you know:: (.) it didn't make much difference because they were out numbered most probably 200 to 1 but .hh no:::: I think the/that's still in the psyche

Res: You think that's still in the psyche?

---

* In all the extracts, *Res* refers to the researcher conducting the interview. The following transcription conventions are used: One or more colons indicate the extension of the previous sound (e.g., Tha::t). Laughter is marked by "hh," and the number of hh's is a rough marker of duration of laughter, while ".hh" indicates audible intake of breath. A question mark is used to mark upward intonation characteristic of a question. Underlining indicates stress placed on a word or part of a word. Extended brackets mark overlap between speakers. The left bracket indicates the beginning of the overlap and the right bracket indicates the end (e.g., hh[hhh] [hhhh]). Numbers in parentheses, e.g. (0.5), indicate pauses in tenths of a second, and (.) indicates a micropause. An equal sign indicates the absence of a discernible gap between the end of a sentence and the beginning of the next sentence. A slash marks a stutter or word correction without a pause, for example "he/he/help." (Inaudible) marks inaudible speech, with time specified in seconds. (Probably) Single parenthesis show guesses of the meaning of hard-to-hear speech.

Joshua: Absolute/it's not long ago (Res: ja/ja) Jeez 1838, (Res: ja/ja) (inaudible) 1837 somewhere around there .hh it's lo:ng/not long ago: in (.) in a:: a:: in the human psyche (Res: ja) you:::

Res: In South African psyche

Joshua: Especially in South African psyche hh where it was (0.5) you know:: um (.) propagated throughout the years you know *die swartes die swartes* [the blacks, the blacks] (Res: ja) and it's (.) they/they can't hide behind anything they were .hh guilty of really, really atrocious in/in those years (Res: ja) of:: um atrocious um (.) barbaric (Res: ja) um::: creeping up to farm houses and burning the whole family inside you know (Res: ja) and (.) u::m ja maybe we don't understand that kind of thing because (0.3) .hh (.) even then the white would never have done that. The white would have gone out (.) you know try to punch the guy:: or (.) you know more ove::rtly aggressive.

▲▼▲

As noted earlier, the social sciences have been predominantly interested in studying beliefs about racial differences with a view to uncovering prejudiced stereotypes or category constructions. Such categorical accounts of blacks were prominent throughout our data corpus. Our white interviewees constructed black beachgoers *inter alia* as frightening, criminal, uncivilized, dirty, and unmannerly. Extracts 1 and 2 show two instances where blacks are described as frightening. Both suggest that black masses have a reputation for violence against whites, and the second attributes this violence to blacks being barbaric.

Without detracting from the importance of this work on stereotyping, we want to suggest that investigations of category construction need to be complemented by investigations of the ontological features of talk, on which the stereotypes are founded. Like all stereotypes, constructions of black aggression and violence are couched in implicit ontological arguments because they attribute such characteristics to *their* nature. However, even as the two extracts show, the ontological features of the talk are not limited to category construction. They are much more pervasive and important. First, the accounts are not only about the Other, but about both *their* nature and *our* nature and, importantly, how these two categories of being have shaped each other. Second, the explanations of difference refer to the forces, processes, and dynamics of human nature that produce, maintain, and shape human difference. In the accounts above, whites' genes, their psychological inheritance, and their natural reactions to fear do the work of maintaining racial separateness.

This ontological talk has features of modern common sense described by Moscovici. The key ontological concepts in the two extracts (i.e., genes and psyche) are the product of modern science. These terms, and theories about biological, psychological and sociological processes, have trickled down into common sense to become the symbolic mediational tools for doing the work of thinking about diversity and difference. Common sense might express a simplified and unelaborated understanding of such concepts, but lay ontological thinking is nonetheless remarkably sophisticated. Both extracts, for example, articulate a complex causal understanding of the impact that a history of racial interaction has had on the biology and psychology of whites living in a postcolonial context.

Earlier, we suggested that the quantitative, correlational studies of explanations of diversity and difference assume that there are distinct explanations, and they emphasize individual differences in preferred explanations. Neither of these assumptions seemed to be supported by the interview data. In the brief passage reported in Extract 1 the speaker talks about culture, inheritance, and genes, alluding to three different bases of racial difference: society, history, and biology. In Extract 2 white resistance to racial contact is theorized as being historically acquired, whereas black aggression is seen to be more racially fixed or endemic. Everyday explanation draws on a multidimensional and complex ontology.

Logical Constraints to Action

Extract 3

Derek: [Explaining the persistence of segregation] I think it's just human nature um certain types of people like to associate w/with certain types I mean you find it amongst white people you find (.).hh some people are conservative some are extravagant some are outspoken and you naturally segre/naturally (.) gravitate to like minded people (Res: ja) so:: .hh and you find religion:: Indian religion and the Christian religion is different so I think (.) I think it's natural that people (.) segregate um to a point but (.) not too artificially you can't (Res: ja) do it artificially no: (Res: ja/ja)

Res: .hh So do you think that um:: um:: (.) forcing segrega::/integration sorry (inaudible)

Derek: Ja forcing integration is not/it doesn't work I'm sure it won't work .hh look I think you force it in terms of schooling and that kind of thing (Res: ja) that/that does help yes but I mean in a private situation like this you can't really .hh force it

Res: Ja (.) ja you can't I suppose=

Derek: =(Even if) artificial structures like putting a boom over the road making people pay and .hh (Res: ja) hopefully (.) people can afford it and other people can't (Res: ja) I don't know if it will work (Res: That's one way) I don't/I don't think you can manipulate those things artificially I think you just allow it (.) to go it's natural course .hh and if you/I mean look at the Berlin wall if try and (.) manipulate it you can't (Res: ja) (.) Eventually it goes one way anyway.

Extract 4

Res: So you mentioned earlier on about culture so do you think that's a cultural thing again? (Jackie: y::es)

Craig: Yes it is definitely (.) I mean this raping of the of the babies was also a culture thing* (Res: ja) This this this u::h this u::h mu::ti doctors or sangomas (.) telling them that if they rape a virgin and the younger the the virgins, the better chance of them getting cured from AIDS (.) I mean that's why they did it (Res: yes) so (.) I mean e::r (.) I I I don't very few white people will believe this (Jackie: they believe anything) not even the Indian will believe that=

Jackie: =you can tell some of them anything they believe it=

Craig: =so but yo::u I mean (.) Zimbabwe one one of their previous elections won that by telling them (.) there's a eye that is watching you when you vote and he will tell Mugabe what you are voting that's why Mugabe=

Jackie: =and they [believe it]

Craig: [inaudible] and (.) I I mean that's (.) I mean that's ridiculous nobody else will believe you except (.) except them and the (.) that's a culture thing (Jackie: ja)=

Res: =and I wonder you can change that then?

Craig: Education but (.) it's taking it will take as I said not for us, it will be (.) maybe the next gener=

Jackie: =our children's children

Craig: ja (.) maybe it will change=

Jackie: =with time (Res: Ja) because their (.) they must be educated...

Craig:... <u>ye::s</u> no I uh uh I think it's a good thing (.) cause that is I think the first step of getting them educated where they will learn that (Jackie: ja) you know it is wrong to throw papers where you sit, I mean it is wrong to u::h (.) you know it is er better to have only one child or or (Res – ja) or or two children if you can't afford it (.) and that there is something like AIDS there is (.) no all these kind of values that they don't learn in the township but (.) that they now can learn and know.

Extracts 3 and 4 paint a different picture of human nature and racial difference than that represented in Extracts 1 and 2. Here the interviewees are speaking as lay anthropologists, theorizing about the cultural features of human nature. This was by far the most prominent kind of lay ontologizing employed by our white participants (Durrheim and Dixon, 2000). Race groups were described as cultural or ethnic groups, and differences between groups were depicted as cultural differences. Culture

* It is a widespread myth in South Africa that sexual intercourse with a virgin will cure AIDS and this had led to a spate of media reports of female infants being raped by HIV infected men.

was spoken about as a set of beliefs, ways of doing things, or tendencies that are associated with, but not necessarily restricted to, historically defined racial groups. For example, in Extract 4 blacks are depicted as a cultural group who rape babies, who believe the myth that such rape is the cure for AIDS, and who have the tendency to believe anything.

Importantly, though, understandings of culture were not simply used as a means of describing categories. Our participants theorized about culture in two related ways, in the process of which they developed lay ontologies of the processes and dynamics of culture. First, they theorized about the causal dynamics — features of basic human nature — that produce and sustain cultures. In Extract 3, for example, we see the common refrain that culture exists because of the tendency of people to associate with their own kind: they "naturally gravitate to like minded people." Second, theories are provided to explain how, as individuals, we get culture, and here lay theories of learning, child rearing, and socialization were common.

In contrast to the essentialism of biological talk about racial genes, cultural characteristics were portrayed as being less monolithic and endemic in race groups and as being open to change. The resources for talking about culture provide theoretical devices for anticipating and explaining (1) differences within cultural and race groups and (2) historical and individual changes in culture. Throughout the interviews, the white interviewees provided enthusiastic support for government-sponsored programs of education, by which it would be possible to change black culture. This provided a framework for rejecting the relevance of race categories in favor of class-based understandings of diversity and difference. As Extract 4 says, a first step toward transformation in South Africa is "getting them educated" (so they will be more like us).

Ontological discourse about human nature served both as a framework for explaining racial difference and as a rationale for racial projects of various kinds. Specific programs of racial action were reasoned to be a logical outcome of facts about the workings of human nature. In contrast to the white Americans of the 1940s, who opposed school desegregation because they believed that blacks were biologically intellectually deficient (Hyman and Sheatsley, 1956), our interviewees supported limited school desegregation because they believed that this would be a way of irradiating racial cultural pathology.

Earlier we noted that the extant research has confirmed that explanations of difference are correlated with prescriptions of racial policy opinions. Our investigation confirms that people's beliefs about the fundamental causes of human difference logically constrain their views about which social policies or interventions are feasible. Reasons for actions and social projects are founded on ontological narratives that map out the bounds of what is possible. Thus, education programs were perceived to have only limited powers for changing culture because — while cultural traits were changeable — segregation itself was a natural cultural process, and racial segregation would thus replicate itself in schools (see Extract 3). Most of the interviewees suggested that, because cultural segregation was an expression of basic human nature, desegregation cannot be forced, and that change would take generations to occur.

Scientific Character of Lay Ontologizing

Extract 5

Fred: You know truthfully I still think it is going even back to the point I made, it's natural segregation. You know, you even look at schools, now the schooling's open and everything like that (.) you still tend to find (0.5) the Indian kids (.) will stick together, the African kids will stick together (Sue: yes) the Portuguese kids. If you got three Portuguese people in the same class, in grade one (.) hell I can almost put my head on a block that those three kids will make friends with each other (Sue: yes) because they have a cultural bond (Sue: hm) uhm and they don't have to know each other from Adam but there's that cultural bond, and you, familiarity, the minute you've got familiarity with somebody that knows your customs and culture.

The theory of social representations suggests that the content of modern common sense is derived from scientific production and retains the objectified character of scientific representation. Such objectified thinking did characterize the explanations of diversity given by our interviewees. Both biological and cultural ontologizing were used to attribute human difference and separation to fundamental human nature. Moreover, specific features and processes of human nature were isolated as explanations. Most commonly, interviewees argued that members of different groups have a cultural bond with each other, feel familiar, comfortable, and relaxed in each other's company, and thus naturally gravitate toward each other, thereby creating patterns of racial segregation and difference (see Extracts 3 and 5).

Durrheim and Dixon (2000) identify a number of features of lay ontologizing that reflect a scientific way of reasoning. Especially important are the objectifying and universalizing character of thinking and reasoning about diversity. We have already considered the objectified nature of these accounts of diversity. Not only are the concepts (e.g., genes, cultural bonds, psyche) scientific objects, but the (causal) processes that dictate historical change and social interaction (e.g., socialization, learning, psychological affinity, and aversion) are concrete processes that have been defined by the various sciences. The language that is used to understand and explain diversity in everyday contexts of social interaction is the objectified language of the sciences.

In addition, explanations of difference were also justified on scientific and empirical grounds. First, universal arguments are put forward to prove the correctness of the objectified theory of difference. For example, the perceived fact that segregation is a natural phenomenon (i.e., an expression of human nature) was believed to be proved by its universal occurrence. According to these arguments, natural segregation stems from processes that apply not only to race relations in South Africa, but equally to race relations elsewhere and even among small children (Extract 5) and to relations between other social categories (religion, culture, nation; see Extract 3). Because segregation occurs everywhere — in different contexts and with different kinds of groups — the reasoning suggests it must be natural and not racist behavior or (in South Africa) a consequence of apartheid.

Rhetorical Structure

EXTRACT 6

Shirley: I got (.) no problem with them if they behave themselves but .hh if they got up and they went home (.) everything is lying down (Res: ja) and/and they just left their/their rubbish there and .hh and (.) as I say to (.) if (.) if you can educate this people (.) (Res: ja) then you can change your beaches but otherwise I can't see:: how you're going to change it (Res: ja/ja/ja)

Res: .hh a number of people that I've spoken to are saying: um similar (.) similar things (.) .hh but how do you change that? Is it education? That's what (.) that's um:: (.) that's what (inaudible) education

Shirley: No:::/no/no:: .hh I don't know if you want my honest opinion you must put off your/your/your tape recorder [hhhh] .hh No I don't think you can change these people (0.5) I don't really think you can (.) Some of them are very::: (.) um civilized but some of them: you can't change them (.) you can't teach them (Res: ja) They're not teach-able (Res: ja/ja/ja) Some of them are (inaudible)

Patrick: Ja::: (.) I don't know um::: if you see their um their ho:mes (.) where they stay in the khayas and things (Res: ja) they (.) they try to clean around it and/and/and use broom and things (.) and paint their houses so (.) [I don't know why]

Shirley: [Some of them ja] absolutely ja (.) ja (Res: ja) I don't think they want to mix as much as (.) us all (.) ja, they don't want to mix with [(inaudible)]

Patrick: [God made us] um different you/you/we/we (Shirley: ja) we/we're white and they're black (Res: ja) So (.) and/and um like sheeps and um/and/and/and/and=

Shirley: =goats=

Patrick: Goats and they (don't) (.) mix (Res: ja) and/and God made us like that (.) so now they (.) all (.) everybody want's us to mix (Res: ja) Why?

Shirley: And if you look at them now=

Patrick: =We're white and they're black (.) why (.) why can't we stay like that? (Res: ja)

Shirley: If you look at them now .hh they'll stay in their own group (.) they don't mix with us (Res: ja) they don't want to (Res: ja) Just as much as we don't want to mix with them (.) they don't want to mix with us=

Patrick: =You get few (.) (fewer than that) (Shirley: ja) want's to make trouble they want to mix here but only for spiteful (Shirley: ja) (Res: ja)

Shirley: Because if you ask their opinion (.) they don't want to come and/come and/and/and sit here and talk to us (Res: ja) We've got nothing in common (0.5) so (.) so why: put us together um let them stay there um I/they don't bother me (.) (inaudible) And we are staying here (Res: ja) and we don't bother them.

At the start of Extract 6 Shirley argues that she has "no problem" with racial mixing, but she ends by echoing apartheid policy, suggesting that blacks and whites should have separate places and should not bother each other by mixing. This flexibility of opinion unfolds in the following sequence of argument. She starts out by suggesting that she has no problem with blacks except for their dirtiness, which she describes as a cultural trait that should be eradicated by means of education. The interviewer then prompts a change in arguments as he inquires whether education really is the key to change. Next, Shirley suggests that education will not entirely resolve racial differences because, by and large, blacks are "unteachable." Patrick immediately rejects this portrayal of innate racial deficiency by arguing that blacks appreciate cleanliness and are a clean people. Then, through joint effort between Shirley and Patrick, we are presented with the argument that, in addition to racial differences being due to cultural or biological factors, they are divinely inspired. Races will keep separate because God didn't make the sheep and goats to mix, and therefore neither blacks nor whites want to mix with each other.

Our earlier analysis has shown how ontological discourse is put to work in contentious contexts where logical arguments must be advanced to defend social practices and where the ontologies themselves must be defended as coherent and empirically grounded. Shirley's about-turn shows how the argumentative context unfolds in conversation and suggests that explanations of difference should be investigated as a flexible rhetorical activity rather than as firm beliefs individuals hold. Even transcendental ontologies — often understood to be the preserve of true believers — were used alongside biological and cultural explanations of difference. As Billig (1988) suggests might be the case, modern common sense also includes transcendental explanations that are not of the objectified, scientific sort.

In order to appreciate the rhetorical advantages of transcendental explanations, it is necessary first to describe some of the features of such an argument and to contrast it with the more scientific-sounding biological and cultural arguments. Arguments about divinely inspired differences did not have the same logical and scientific character as objectified common sense, and they were presented as uncontested and incontestable truths. In contrast to material categories and processes, transcendental agency is abstract and invisible, and accounts of transcendental causality were thus limited to assertions that "God made it so." These assertions were supported by claims that sheep and goats don't mix — nor do other animals like horses and donkeys, birds and bees, shark and fish, *ad nauseam*. Although such gestures mimic evidence of the hand of God, they were simply claims that the divine and abstract principle of separation is expressed throughout nature. Such transcendental arguments, therefore, do not anticipate change but see the same abstract and static transcendental principle in all nature and in all time. Unlike biological and cultural explanations, therefore, transcendental explanations of diversity made no references to processes that produce historical change and social interaction.

Despite these differences, transcendental explanations of diversity also portrayed difference as natural and could thus hitch objectified explanations of process onto the transcendental account. Accordingly, although Shirley and Patrick agree that God made the races different and instilled in them a desire to remain separate, they also argued that this divinely inspired human nature manifests itself psychologically in a preference for one's own group and a desire to remain apart. The reason why segregation is preferred, although ultimately attributable to God, is because of the mundane psychological fact that neither blacks nor whites really want to mix with each other, because they have nothing in common.

Each kind of ontological argument allows speakers to do different things. They each convey special rhetorical advantages and thus find their own place in the flexible repertoire of explaining racial diversity. Cultural arguments permit the speaker to anticipate social change, accept class-based racial convergence, and avoid accusations of racism. Biological arguments allow speakers to construct racial differences that are not amenable to change and thereby explain the limits to racial convergence. However, such arguments are susceptible to critical empirical refutations of the categorical nature of racial difference, and they are thus open to criticisms of racism. Transcendental arguments also facilitate arguments against racial convergence and integration, but because they are presented as incontestable truths, they are not susceptible to criticisms regarding the veracity of their details. The abstractness of transcendental arguments bestows important rhetorical advantages.

Qualitative studies of lay ontologizing suggest that there is no uniform typology of explanations of human diversity that match specific categories of people. Individuals use variable repertoires of explanation in accounting for segregation. Thus, the individual — the fulcrum of a correlational analysis — cannot act as the unit of analysis for such investigations. On the contrary, the focus of investigation should be on the conversational context of justification and criticism in which lay ontologies are presented. As discussions of race unfold, the changing interactional context will prompt variability in ontologizing as interlocutors adjust their thinking to the rhetorical demands of the conversational context. All this suggests that we need to study the demands of the conversational context to understand the utility of different kinds of lay ontologizing. Biological, cultural, and transcendental ontologies each convey special rhetorical advantages, allowing speakers to justify and criticize different aspects of social policy.

Political Functions

EXTRACT 7

Marvin: =Why is only the sardines coming here? (Res: ja) Why? (2) Because they like the company of other sardines (Res: ja (.) ja/ja) Not (inaudible) crayfish and (.) sharks they are together because they belo::ng together (Res: ja) You and me can do nothing about it the black man wants to be with the black people (Res: ja) It's got nothing to do with racism (Res: ja) because he enjoys his own people. He knows them. They know him (Res: ja) It's because of different cultures, different allegiance, different lots of things (Res: ja) (.) That-is-why-the-black-man-likes-to-be-with-his-people (*measured)* (Res: ja) He's not happy if he has to come and sit here and (.) (inaudible) white people sitting here (Res: ja) The same as you won't (.) you'll be lonely .hh if there's 500 black people sitting there and there's white/white man/you'll come and sit here .hh you'll be lonely (Res: ja/ja) It's a matter you:: (.) want to be with your own (.) I mean that's natural (Res: ja) the lion goes with the lion (Res; Ja) and the springbuck goes with the springbuck (.) Why would you (.) go one/one springbuck go in between 50 lions (Res: ja) or the other way around.

Res: So you think that um the reason why there's still separation=

Marvin: =Tha::t's natural you won't change it. You can't change it my friend. The Hollanders will be with the Hollanders (.) the Germans will be with the Germans (Res: ja) and the Jews will be with the Jews .hh (inaudible) (Res: ja) and for (.) for everything it is like that it's natural.

Since the 1970s social scientists have argued that new forms of subtle racism have emerged to replace biological racism in Western societies (Sears, 1988). In contrast to the crude biological beliefs of old-fashioned racism, subtle racism draws on notions of culture to understand human difference and to inform racial projects (Pettigrew and Meertens, 1995, Bobo et al., 1997). Partially because of this shift away from biological discourse, these newer forms of racial expression are seen to be less offensive than older blatant manifestations ("societal discourse about racial groups has been substantially cleaned up" [Sears, 1998: 98]), and indeed, some have even suggested that they might not be racist at all (e.g., D'Souza, 1995; Roth, 1994).

Our analysis suggests that there are important differences between the kinds of racial projects that can be logically grounded in biological and cultural ontologies. Biological accounts of inherent inferiority prescribe monolithic programs of racial segregation and unequal treatment of "unteachable" racial bodies. In contrast, cultural discourse anticipates the educability of individuals and thus provides a rationale for class-based racial projects, informed by the values of egalitarianism, individualism, and fairness.

Nevertheless, our investigation of lay ontologizing suggests that there may be less difference between biological and cultural discourse (and even transcendental discourse) than the current literature assumes. These forms of ontologizing perform very similar political functions as they transform racial projects — both collective and personal — into dictates of nature. Extracts 1 to 7 show that our interviewees were able to provide similar defenses of racial segregation using biological, cultural, and transcendental ontologizing. In all instances, the crucial hinge of the argument was the naturalness of segregation, which served to deflect criticisms of racism. As Marvin says, his opposition to racial integration has got "nothing to do with racism." Rather, it is an expression of human nature, and there is "nothing you can do about it" (Extract 7). Politically, it doesn't matter so much whether we evolved (or were created hardwired) to segregate or whether segregation is a cultural inclination. What matters is that in attributing segregation to nature, we can universalize our preferences and render the racial project necessary. We can thus justify the racial project, deflect criticisms of racism, and criticize any "irrational" or "politically-inspired" attempt to challenge the racial project, arguing that it is doomed to failure because "you can't change what's natural" (Extract 6).

Ontological discourse grounds racial projects in a philosophy of history, or, according to Balibar, in a

> *histriosophy* which makes history the consequence of a hidden secret revealed to men [sic] about their own nature and their own birth. It is a philosophy which makes visible the invisible cause of the fate of all societies and people; not to know that cause is seen as evidence of degeneracy or of the historical power of evil. (Balibar, 1991a:55)

Lay ontologizing about human diversity performs political and ideological functions, reducing political projects to natural necessities. In our interviews, cultural, biological, and transcendental ontologizing were all used in a reactionary politics opposed to social transformation and desegregation in South Africa. This was done by colonizing nature, and by placing limits on what is and what is possible (Therborn, 1980) in the field of social relations.

HISTORICAL AND SOCIAL BASIS

> It is "common sense" [to white South Africans] that black people are inferior to white people . . .. They see black "inferiority" as one of the imperatives of human nature. They "explain" a social fact by direct reference to biology and thereby misunderstand it. If we assume the natural inferiority of a group we don't look for social causes of its actual "inferiority". If, on the other hand, we discover that there is no biological root for this "inferiority", no imperative of human nature, then we begin to ask the illuminating question: What is the social structure that creates the various objective "inferiorities" of certain groups? And only then do we understand how our society operates. (Turner, 1972 cited in Maré, 2002:11).

In the middle years of apartheid rule, Rick Turner observed that biological explanations of black inferiority were common sense among white South Africans. This is no longer the case 30 years later, but common sense did not change in the way Turner anticipated. Biological discourse and the policy of separate development were not replaced by a radical social structural discourse and a program for changing the way in which society operates. Instead, cultural explanations of differences predominate today, often used in tandem with biological and transcendental arguments. This historical change in the everyday explanation of racial difference follows a similar pattern to that which occurred with the change from Jim Crow racism to laissez faire racism in the United States (Bobo et al., 1997). In Europe, too, Balibar (1991b) argues that the dominant theme of racism "is not biological heredity but the insurmountability of cultural differences, a racism which, at first sight, does not postulate the superiority of certain groups or peoples in relation to others but 'only' the harmfulness of abolishing frontiers, the incompatibility of life-styles and traditions" (Balibar, 1991b:21).

We have argued that cultural discourse performs political and ideological functions similar to biological discourse, naturalizing racism and transforming it into a necessity and an eventuality. In addition we have suggested that cultural discourse confers particular rhetorical advantages, allowing speakers to deflect accusations of racism while at the same time opposing social transformation.

Cultural discourse serves the requirements of contemporary racism better then biological discourse for a number of reasons. First, because cultural discourse anticipates the possibility of individual change, it is better aligned with the contemporary liberal values of individualism and equality. In contrast to the monolithic segregation recommended by biological discourse of endemic racial inferiority, cultural discourse rationalizes flexible, class-based systems of segregation, incorporating assimilation and social mobility (Goldberg, 1998).

In addition, cultural discourse is more practical for regulating behavior in desegregated contexts of contemporary liberal democracies. Biological discourse draws heavily on a sexual metaphor — genetic transfer, heritability, breeding — to explain how contact produces negative (unnatural) consequences. Cultural discourse, in contrast, employs social metaphors of contact — assimilation and mixing — to explain how integration produces negative outcomes (discomfort, fear, etc.). This makes cultural discourse more useful for arguing and thinking about "race" in myriad local situations of impending social and racial contact that present themselves on a daily basis in the new South Africa and other "open" societies. In moving from a sexual to a social metaphor of contact, cultural discourse has racist utility in widening the range of sites and contexts where naturalizing discourse can function plausibly to warrant segregation.

CONCLUSION

Concerns about the ethics of scientific production in disciplines such as biology, anthropology, and psychology often arise when scientific thinking is taken up and put to use in society at large by governors, administrators, and citizens. Reflections on the ethics of scientific endeavors should thus extend outside of the context of scientific discovery, justification, and publication. In addition to asking whether scientists are behaving ethically, we should also be asking how society at large is making use of scientific productions. These are the kinds of ethical questions that Einstein began asking later in his career when he realized how his work in nuclear physics was being put to use in the production of nuclear weapons that "threaten the continued existence of mankind."*

Following Moscovici, we have argued that scientific production becomes socially efficacious in the process of being popularized and in becoming part of common sense. Modern common sense is composed of simplified versions of the concepts, ideas, and theories of science, which provide representations of the content of nature. These representations then inform and structure collective and individual racial projects of various kinds, for they provide an ontological grounding for what is and what is possible.

The primary aim of this chapter has been to describe the explanations of racial difference provided by ordinary people. Using extracts from interviews about desegregation with white South Africans, we characterized racial explanations in terms of six features:

1. The explanations are grounded in complex multidimensional ontologies that draw on biological, cultural, and transcendental discourse, and that provide not only a description of social categories but also an account of the processes that produce and regulate racial difference.
2. In specifying what is, these ontologies also prescribe what is possible and thus provide logical constraints for action.
3. The truth of the ontologies and the validity of ontological reasoning are grounded in empirical and theoretical arguments that have a scientific character.
4. Ontologizing is sensitive to the rhetorical demands of conversational context, and therefore speakers use a flexible repertoire of ontological argument.
5. Ontological discourse serves political functions in justifying and defending racism and criticizing social transformation.
6. Explanations of difference change historically, adapting themselves to the rhetorical, political, and material demands of the social context and changes in scientific truths.

Does it follow that a change in the politics of the everyday use of scientific theorizing will emerge after a change in the content of scientific theorizing? Can scientists adjust their activity, given their view of the use

* Russell-Einstein manifesto, London, 9 July 1955; http://www.pugwash.org/about/manifesto.htm, accessed January 25, 2006.

of their work in grounding racial projects of various kinds? Our investigation suggests that cultural discourse serves similar political functions as biological and transcendental discourse. However, what is common about these kinds of ontologies is that they reify human essence in ignoring the impact of the historical and social forces that have produced racial diversity. What seems to be missing from the everyday explanations is a view about how social structure creates racial difference. Such a scientific project would certainly provide the basis for a critical common sense.

REFERENCES

Apostle, R.A., Glock, C.M., Piazza, T., et al., *The Anatomy of Racial Attitudes*, University of California Press, Berkeley, 1983.

Balibar, E., Racism and nationalism, In *Race, Nation, Class: Ambiguous Identities*, Balibar, E. and Wallerstein, I., Eds., Verso, London, 1991a; chap. 3, pp. 37–67.

Balibar, E., Is there a neo-racism?, In *Race, Nation, Class: Ambiguous Identities*, Balibar, E. and Wallerstein, I., Eds., Verso, London, 1991b; chap. 1, pp. 17–28.

Billig, M., Social representation, objectification and anchoring: a rhetorical approach, *Social Behaviour*, 3, 1–16, 1988.

Billig, M., *Arguing and Thinking: A Rhetorical Approach to Social Psychology*, 2nd ed., Cambridge University Press, Cambridge, 1996.

Bobo, L. and Kluegel, J.R., Opposition to race-targeting: self-interest, stratification ideology or racial attitudes? *American Sociological Review*, 58, 443–464, 1993.

Bobo, L., Kluegel, J.R., and Smith, R.A., Laissez faire racism: the crystallization of a kinder gentler anti-black ideology, In *Racial Attitudes in the 1990s: Continuity and Change*, Tuch, S.A. and Martin, J.K., Eds, Praeger, Westport, CN, 1997.

Danziger, K., The methodological imperative in psychology, *Philosophy of the Social Sciences*, 16, 1–13, 1985.

De Certeau, M., *The Practice of Everyday Life*, Rendall, S., Trans., University of California Press, Berkeley, 1984.

Dixon, J. and Durrheim, K., Contact and the ecology of racial division: some varieties of informal segregation, *British Journal of Social Psychology*, 42, 1–24, 2003.

D'Souza, D., *The End of Racism*, Free Press, New York, 1995.

Durrheim, K. and Dixon, J., Theories of culture in racist discourse, *Race and Society*, 3, 93–109 2000.

Durrheim, K. and Dixon, J., The role of place and metaphor in racial exclusion: South Africa's beaches as sites of shifting racialization, *Ethnic and Racial Studies*, 24, 433–450, 2001.

Durrheim, K. and Dixon, J.A., *Racial Encounter: The Social Psychology of Contact and Desegregation*. Psychology Press, London, 2005.

Flick, U., Everyday knowledge in social psychology, In *The Psychology of the Social*, Flick, U., Ed., Cambridge University Press, Cambridge, 1998.

Goldberg, D.T., The new segregationism, *Race and Society*, 1, 15–32, 1998.

Hyman, H.H. and Sheatsley, P.B., Attitudes towards desegregation, *Scientific American*, 195, 16–23, 1956.

Kluegel, J.R., "If there isn't a problem you don't need a solution": the bases of contemporary affirmative action attitudes, *American Behavioral Scientist*, 28, 761–784, 1985.

Kluegel, J.R., Trends in whites' explanations of the black-white gap in socio-economic status, 1977-1989, *American Sociological Review*, 55, 512–525, 1990.

Kluegel, J.R. and Smith, E.R., Affirmative action attitudes: effects of self-interest, racial affect and stratification beliefs on whites' views, *Social Forces*, 61, 797–824, 1982.

Kluegel, J.R. and Smith, E.R., Whites' beliefs about blacks' opportunity, *American Sociological Review*, 47, 518–532, 1982.

Kluegel, J.R. and Smith, E.R., *Beliefs About Inequality: Americans' Views About What Is and What Ought To Be*, Aldine de Gruyter, Hawthorne, NY, 1986.

Maré, G., Thinking about Race and Race Thinking, Inaugural Lecture, University of Natal, Durban, South Africa, September 2002.

Moscovici, S., Foreword, In *Health and Illness: A Social Psychological Analysis*, Herzlich, C., Ed., Academic Press, London, 1973.

Moscovici, S., *La Psychanalyse: Son Image et son Public*, Presses Universitaires de France, Paris, 1976.

Moscovici, S., The phenomenon of social representations, In *Social Representations*, Moscovici, S. and Farr, R., Eds., Cambridge University Press, Cambridge, 1984a.

Moscovici, S., The myth of the lonely paradigm, *Social Research*, 51, 939–967, 1984b.

Moscovici, S., Notes towards a description of social representations, *European Journal of Social Psychology*, 18, 211–250, 1988.

Moscovici, S., The history and actuality of social representations, In *The Psychology of the Social*, Flick, U., Ed., Cambridge University Press, Cambridge, 1998.

Omi, M. and Winant, H., *Racial Formation in the United States: From the 1960s to the 1990s*, 2nd ed., Routledge, New York, 1994.

Pettigrew, T.F., Racial change and social policy, *Annals of the American Academy of Political and Social Science*, 441, 114–131, 1979.

Pettigrew, T.F. and Meertens, R.W., Subtle and blatant prejudice in Western Europe, *European Journal of Social Psychology*, 25, 57–75, 1995.

Potter, J. and Wetherell, M., *Discourse and Social Psychology: Beyond Attitudes and Behaviour*, Sage, London, 1987.

Potter, J. and Wetherell, M., Social representations, discourse analysis and racism, in *The Psychology of the Social*, Flick, U., Ed., Cambridge University Press, Cambridge, 1998.

Roth, B.M., *Prescription for Failure: Race Relations in the Age of Social Science*, Transaction, New Brunswick, NJ, 1994.

Samelson, F., From "race psychology" to "studies in prejudice": some observations on the thematic reversal in social psychology, *Journal of the History of the Behavioural Sciences*, 14, 265–278, 1978.

Sears, D.O., Symbolic racism, In *Eliminating Racism: Profiles in Controversy*, Katz, P.A. and Taylor, D.A., Eds., Plenum, New York, 1988.

Sears, D.O., Racism and politics in the United States, In *Confronting Racism: The Problem and the Response*, Eberhards, J.L. and Fiske, S.T., Eds., Sage, Thousand Oaks, CA, 1998.

Schütz, A., *Collected Papers*, Vols. I–III, Nijhoff, Den Haag, The Netherlands, 1966.

Therborn, G., *The Ideology of Power and the Power of Ideology*, Verso, London, 1980.

PART 2

GENETICIZATION AND THE NATURE OF DIFFERENCE

5

INVENTING THE HISTORY OF A GENETIC DISORDER: THE CASE OF HUNTINGTON'S DISEASE

Alice Wexler

CONTENTS

WITCHES, CRIMINALS, AND HUNTINGTON'S DISEASE

In 1932 an American psychiatrist named Percy R. Vessie, medical director at an elite private sanitarium in Connecticut, published a paper about New England witches and ne'er do wells that became an unofficial origins story of Huntington's disease in the United States (Vessie, 1932). Formerly called Huntington's chorea on account of the jerky, involuntary movements called chorea that form its most dramatic symptom, this disorder is a dominantly inherited neurological illness also characterized by personality changes and cognitive decline, leading inexorably to dementia and death over a period of 15 to 20 years (Harper, 1991).* In 1872 a young North American physician, George Huntington, wrote the classic description, which attracted the attention of eugenicists such as Charles B. Davenport after the turn of the century (Huntington, 1872). Davenport, in turn, commissioned Elizabeth B. Muncey to carry out the first large family field study of the disease under the aegis of his newly established Eugenics Record Office (ERO). Vessie drew on Muncey's data when he wrote his own paper in 1932, by which time Huntington's chorea had become widely known within the international neurological community.

Starting with one of his patients, Vessie constructed a pedigree (see Figure 5.1) dating back to three seventeenth-century men and their wives, all of whom, according to Vessie, had emigrated to New England from the East Anglian village of Bures, in Suffolk, in the 1630s, thereby establishing in the United States, as he put it, "a chain of generations" that showed "no disposition to a cessation of this horrible heredity" (Vessie, 1932:565). For Vessie, the men's alleged social pathology and acts of delinquency signaled that they may have suffered from Huntington's chorea. As thieves, quacks, and lawbreakers, they were "notorious principals in unsavoury colonial history" (Vessie, 1932:556). In addition, they were "illiterate and arrogant, and none attained recognition or respectability" (Vessie, 1932:73). Vessie's portrait of their descendants was equally pejorative, noting that "their spitefulness, bickering, querulousness and blasphemy bring contempt, dislike and hatred to their doors; therefore, the pity they would receive because of their bizarre and dancelike movements is in many cases prevented by their cantankerous behaviour" (Vessie, 1932:572). Their insistence on marrying cousins and refusal to stop having children was a sign of their "persistent stupidity," which "contributed to the dooming of many of their descendants" (Vessie, 1932:565).

* Huntington's disease affects both men and women, with symptoms developing gradually in mid-life, most often in the thirties or forties. Those with an afflicted parent have a 50–50 risk of inheriting the disorder, since a single copy of the dominant genetic variant (i.e., allele) causes the disease. Moreover, with close to 100% penetrance, this gene presents offspring with a stark situation: if they inherit the gene they will almost certainly develop the disease and pass on a similar risk to their own children. If they do not develop symptoms, they will not pass on the disease. In the past, in communities where affected families lived, Huntington's was popularly called St. Anthony's dance, magrums, or St. Vitus's dance, though the latter term more accurately refers to childhood or Sydenham's chorea, a self-limiting disorder often following rheumatic fever. Since Vessie used the name Huntington's chorea, I will follow his usage when referring to his analysis; otherwise I will use the current term, Huntington's disease.

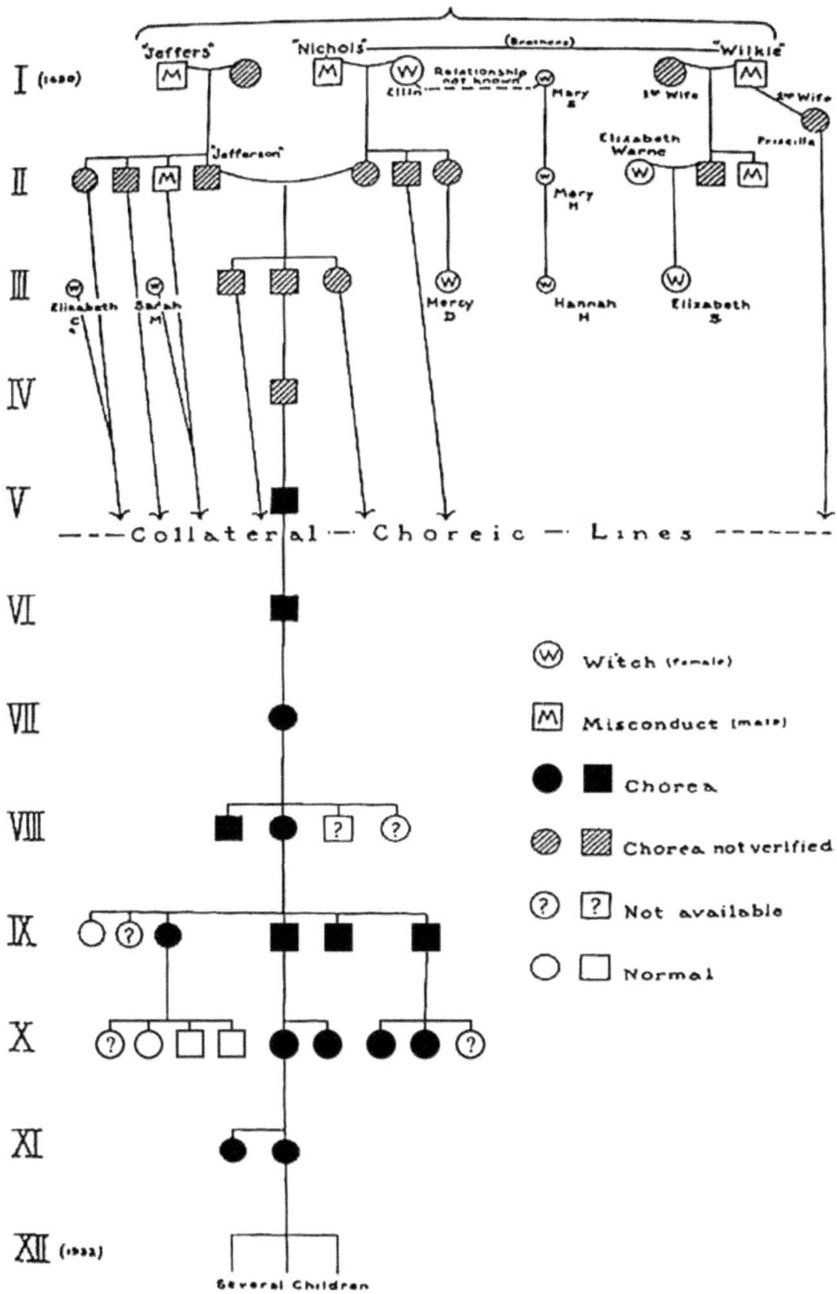

Figure 5.1 Pedigree chart of the Bures family group showing members tried in colonial courts for witchcraft and misconduct and the direct descent of Huntington's chorea in later generations (Vessie, 1932:554).

What especially captured Vessie's imagination were the alleged associations with witchcraft of the Bures wives — or specifically one of the wives and some of her supposed female relatives. According to Vessie, "Not only do the archives disclose behaviour problems in these men, but we suspect their wives of being the lamentable means of transporting a family disease from England to the colonial states, which inheritance has spread throughout the United States. We believe the true story of this lesion to be revealed in the witchcraft trials of women in the Bures group" (Vessie, 1932:573). In other words, the accusations of witchcraft against the women, like the legal problems of the men, were markers signaling the presence of the disease. In 1939, in a second article, Vessie further denounced the "bad characters from Bures." In his view, the eugenic lesson of this story was clear. The "sorrowful march of victims to the scaffold, exile and social ostracism" dramatized the urgent need for "rigid sterilization" (Vessie, 1939:600). Though most of Vessie's genealogical claims were discredited in the late 1960s (Hans and Gilmore, 1969), his story is still cited today (Aronson, 2002).

TEXTUAL BODIES AND REAL LIVES

In *Extraordinary Bodies: Figuring Physical Disability in American Culture and Literature*, Rosemarie Garland Thomson (1997: 22) describes the body as "a cultural text that is interpreted, inscribed with meaning — indeed made — within social relations." According to Thomson, visual and verbal images often assign stereotyped negative meanings — inferiority, ugliness, abnormality, stupidity, dirtiness, or danger — to visible differences in ways that obscure the lives of real disabled people. Such meanings also inform "the identity — and often the fate — of real people with extraordinary bodies." As Paula Treichler has written of AIDS, "our names and representations . . . influence our cultural relationship to the disease and, indeed, its present and future course" (Treichler, 1988:95).

Medical narratives form one genre of such representations, shaping images of the ill and of disabled persons who become patients when they seek medical help. I suggest that Vessie's sensationalized tale linking witches and ne'er-do-wells with people with Huntington's chorea influenced not only the clinical practice of physicians but also the research agendas of scientists and the self-images and identities of affected families, in ways that we are still struggling against today. Of course there is nothing intrinsically stigmatizing about the claim that certain seventeenth century women accused of witchcraft or men who had run-ins with the law may have suffered from symptoms resembling those of present-day Huntington's disease. The idea that involuntary movements and mental changes similar to those of this disease might sometimes have aroused suspicions of supernatural activity does not cast aspersions on the sufferers, but Vessie's peculiar and prurient accounts of witchcraft, juxtaposed with hostile and pejorative accounts of people with Huntington's, denigrated both seventeenth century female witches and twentieth century sufferers from chorea, further associating the disease with danger and destructiveness and linking women with witchcraft and disease. Vessie wrote,

> The devil visits the witch at night to consort with her, claws and scratches her, leaving conspicuous marks upon her skin. She usually develops nipples for imps to suckle. They attach themselves to her body, usually hanging from the breasts and genitals. The witch is accused of calling the imps by name, such as Tom, Will, or Kit, and they were dispatched to obey her commands at her pleasure, placing a curse on enemies, causing them to die, making children in the neighbourhood cry in fear, producing strange fits in innocent persons, killing hogs and cattle and producing storms. Caught in this web, she is finally forced to submit to every conceivable temptation: stealing, murder and even suicide." (Vessie, 1932:560)

The witches' alleged descendants afflicted with Huntington's chorea were equally repellent, according to Vessie. "So terrifying to observers were the repulsive aspect and ferocious irritability of its victims," he wrote, "that parents in New England down to the present day teach their children to fear, hate and shun these 'living examples of sin'" (Vessie, 1939:599).

HISTORICAL CONTEXTS

To grasp the cultural import of Vessie's narrative, it is important to note the historical contexts in which his two papers appeared. In 1927, the *Buck vs. Bell* Supreme Court case in the United States, upholding Virginia's involuntary sterilization law in the case of a young woman named Carrie Buck, had "greatly boosted the pace at which sterilization programs were enacted and implemented" according to historian Philip Reilly (1991:87). By 1932 some 16,066 institutionalized individuals in the United States had been sterilized. The following year the Nazi Eugenic Sterilization Law was enacted in Germany, mandating compulsory sterilization for those with Huntington's chorea, as well as people with other specified diseases and disabilities. By the end of 1937 some 225,000 people in Germany had been sterilized. This policy was well known in the United States and even approved of by leaders of the mainline eugenics movement, whom the Nazis cited as the inspiration for their legislation. In 1939, the year in which Vessie's second paper appeared, the Third Reich moved toward euthanasia of the mentally ill and disabled. In the United States by that year, the number of people involuntarily sterilized had risen to 33,035, nearly two-thirds of them women.

Indeed, the 1930s were the "high water mark of eugenic sterilization in the United States" (Reilly, 1991:128). A number of states had laws prohibiting the marriage of people with epilepsy, "hereditary insanity," and other conditions deemed inherited. Even the 1935 Myerson Report of the American Neurological Association, "the definitive scientific critique of eugenic sterilization published in the United States," according to Reilly (1991:123), recommended sterilization both inside and outside of institutions in cases of certain clearly inherited diseases, such as Huntington's chorea, since "the indications for sterilization are usually obvious to the physician and should be so to the patient" (Myerson et al., 1936:180). According to Reilly (1991:87), the years from 1927 to 1942 in the United States were a "triumphant period for those who embraced hereditarian hopes for social progress." Though sterilization lost adherents in the post-World War II period, the practice continued, with at least 60 thousand people in the United States sterilized by the 1960s (Reilly, 1991). Vessie's call for "rigid sterilization," then, was not abstract speculation but support for a policy that was already in practice.

It is also worth noting that Vessie's 1932 article appeared not in an obscure publication, but in the unofficial organ of the American Neurological Association, the *Journal of Nervous and Mental Diseases*, edited in New York by the respected Smith Ely Jelliffe. It immediately attracted notice in the prestigious British medical journal, the *Lancet*, which praised it as a "fascinating study" (*Lancet*, 1933:1869–1870). Indeed, the *Lancet* believed that Britons "may congratulate ourselves" on the loss of the three Bures couples to America, since they and their progeny were clearly "undesirable characters" who would nowadays be classed "as belonging to the social problem group" (*Lancet*, 1933:870). The popular U.S. magazine, the *Literary Digest* also picked up the story, republishing the *Lancet* abstract under the title "The Witchcraft Disease" (*Literary Digest*, 1933).

In 1934, an up-and-coming English neurologist, MacDonald Critchley, published his own elaboration of Vessie's thesis in the *British Journal of State Medicine*, claiming to identify the English mother of the three men of Bures, a woman whom he flippantly referred to as "the gay lady of Bures" (Critchley, 1934:587) and subsequently as "the wanton Mary Haste" (Critchley, 1964:215) since her supposed offspring was a "natural son." For Critchley, not only the affected individuals but also their unaffected relatives were "liable to bear the marks of a grossly psychopathic taint, and the story of feeblemindedness, insanity, suicide, criminality, alcoholism and drug addiction, becomes unfolded over and over again" (Critchley, 1934:575). Within the neurological and genetic literature on Huntington's chorea, Vessie's and Critchley's papers soon became "the two most quoted classics" (Caro and Haines, 1975:91–95). Vessie's thesis even entered popular culture, in the British 1988 mystery novel *The House of Stairs*, with a narrator at risk of Huntington's whose ancestor "from far, far back, was a Bures woman" (Vine, 1988:93).

HUNTINGTON'S CHOREA AS HISTORY AND MYTH

How valid were Vessie's claims as history — by the historiographical standards of his own time and by the understandings of our own? On closer inspection, the alleged "misconduct" of the men and women from Bures turns out to be the ordinary behavior of laborers or yeoman in the Puritan colonies: selling beer without a license, making a speech against the governor, signing harsh verdicts when serving on a jury, paying a fine for "retaining a stolen calf," running a tavern, selling medicine that was initially deemed useless for scurvy (a judgment later overturned), paying fines for unspecified infractions, and "distemper" (drunkenness) (Vessie, 1939:556–560). The Puritans were a litigious people, and many individuals ended up in court for one reason or another (Bushman, 1967). These infractions were probably indicators of class and in some instances, political protest, not chorea.*

Vessie's genealogical and historical claims about witchcraft also present problems. His most glaring error was the claim that Elinor, the wife of Nicholas Knapp, one of the Bures men, was executed as a witch. As Hans and Gilmore (1969) point out, Vessie conflated the wife of Nicholas Knapp, Elinor, with Goodwife Knapp, a convicted witch who was hanged in Fairfield, Connecticut, in 1653 and had no relationship to Vessie's pedigree. Elinor and Nicholas Knapp did indeed count families with Huntington's chorea among their later descendants, but according to sources used by Vessie, Elinor Knapp died in Stamford, Connecticut, in 1658, five years after the convicted witch, and there is no evidence that Elinor was ever accused of witchcraft (Hans and Gilmore, 1969).

Nonetheless, Vessie attempted to link her with several other accused witches, implying a family connection and speculating that all of them may have been afflicted with chorea. In one case, the mere fact — "strange to note" — that a mother, daughter, and granddaughter were all accused of witchcraft suggested to Vessie that they were all affected with the disease, particularly since one witness to their trial alluded to "strange

* One of Vessie's three men — who was not from Bures — became a leading figure in several town settlements, was elected freeman, and was a holder of local office, with descendants who played roles of distinction in the later history of Connecticut (Mead, 1911).

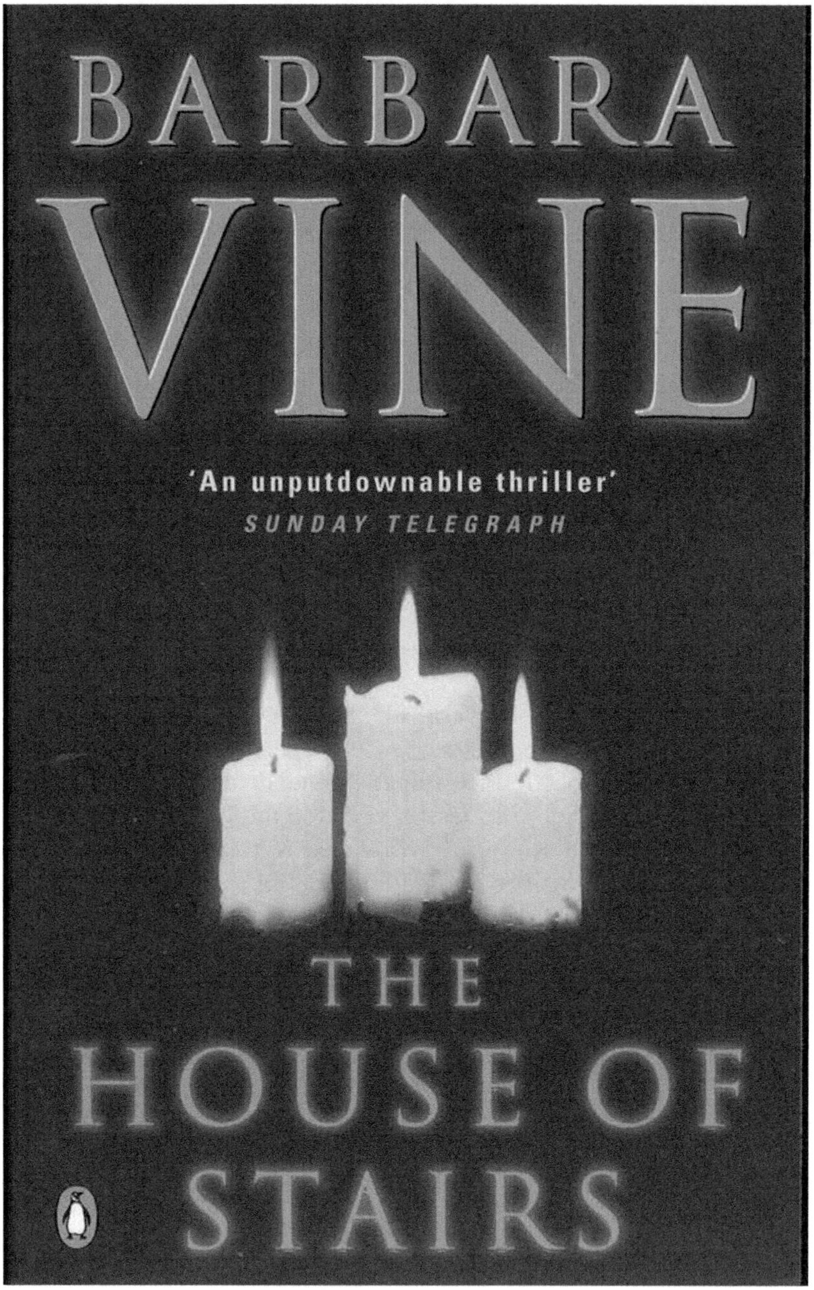

Figure 5.2 The fictionalization of Vessie's (1932) witchcraft thesis in the 1988 novel *The House of Stairs* by Barbara Vine (aka Ruth Rendell) in which a narrator had an ancestor "from far, far back, [who] was a Bures woman" (1988:93).

gymnastic performances" (Vessie, 1932:562). He implied that another accused woman, Elizabeth Clawson, with no known relation to any of the later affected families or to the Bures group, also supposedly suffered from chorea. The "violent motions" and "agitations" (Paige, 1877: 652) of "the celebrated Groton witch" Elizabeth Knapp (the granddaughter of a second Bures man, William Knapp) also reminded Vessie of Huntington's, even though sixteen-year-old Elizabeth Knapp's "fits" began suddenly and lasted for about three months, after which she evidently recovered and went on to marry and live out her life as a normal New England wife and mother (Vessie, 1932:557). Indeed, the Groton episode, which occurred in 1672, was a famous and well-documented case of possession that did not resonate at all with descriptions of Huntington's chorea. In fact, all of the suspected witches named by Vessie are well known to scholars, and none of them showed behaviors reminiscent of this disease.

As Hans and Gilmore (1969) pointed out, Vessie built layer upon layer of erroneous genealogies to construct a misleading portrait that implied a connection between many of the accused witches both with each other and with later descendants with Huntington's chorea. It is true that Vessie relied on faulty genealogical and diagnostic information in the unpublished pedigree of Elizabeth Muncey, the physician and eugenics worker who had carried out the field study of Huntington's chorea families in 1912 and 1913 under the aegis of the ERO and Charles B. Davenport. It was Muncey who first conflated Elinor Knapp with Goodwife Knapp, claiming erroneously in her field notes that Elinor, the wife of Nicholas Knapp, was hanged as a witch in 1658. Muncey also "inferred" that Elizabeth Knapp had suffered from Huntington's chorea (Muncey, 1913). Charles Davenport, too, had speculated in his influential published analysis of the Muncey material that "in some cases the [genealogical] chain is partly completed in the remoter ancestry by the history of a case of witchcraft. It is not strange that at a time when many nervous disorders were ascribed to witchcraft, it should be thought that a choreic was bewitched or even capable of bewitching others — especially her own children!" (Davenport, 1916:202).

Moreover, while much recent historical research makes clear that early New Englanders had a naturalistic understanding of mental illness — often called "distraction" or "lunacy" — the historiography of the 1930s was far more likely to accept the popular notion "that witches have usually been deranged persons, insane or at least deeply eccentric" (Demos, 1982:90). Vessie's claim that "certain nervous and mental disorders must have been regarded with consternation and horror, as the workings of the devil" was not so far fetched given the sources available to him. At the very least, Huntington's chorea sometimes produced behaviors (extreme irritability, depression, outbursts of rage, paranoia, and other personality changes) that even present-day historians have ascribed to those accused of witchcraft.

Carol Carlson's (1987:118) portrait of the accused witch as "an angry, assertive, ill-tempered, quarrelsome, and spiteful" woman does resonate with symptoms of Huntington's disease in some individuals (Demos, 1982).

However, it is more likely that involuntary bodily movements and strange behavior would have been interpreted as signs of possession or as an indication that someone had been bewitched and was, therefore, not accountable for her behavior. According to the historian Michael MacDonald (1981: 209), "jerking helplessly like a marionette controlled by a careless puppeteer was to the popular mind a strong reason to suspect witchcraft," but the sufferer was considered a victim of witchcraft, not a witch herself. The distinction between a witch and a person bewitched or possessed by the devil (a demoniac) was the difference between guilt and innocence. The devil was not "inside a witch's body, as he is in a demoniac's; in consequence, a witch does not suffer from convulsions and a demoniac does. A witch has voluntarily entered into association with a devil, whereas possession is involuntary and a demoniac is not, therefore, responsible for her wicked actions, as is a witch" (Walker, 1981:10).* Though witchcraft persecutions declined in the late seventeenth century, as late as 1769, William Buchan's popular medical text, *Domestic Medicine*, defined St. Vitus's dance as a "particular species of convulsion fits . . . wherein the patient is agitated with strange motions and gesticulations, which by the common people are generally believed to be the effects of witchcraft" (Buchan, 1985:551–552).

Were Vessie's misleading claims merely a result of reliance on mistaken genealogies, careless research, and assumptions about New England culture prevalent in the 1930s? Considering that he had access to more accurate genealogies, which he even cited in his paper, his genealogical distortions seem more likely to have grown out of his prior beliefs, even prejudices, about Huntington's chorea. These beliefs could easily have obscured the more truthful information available to him and made links between the disease and witches appear reasonable and plausible. After all, Vessie lived and worked near the town of Stamford, where affected families lived. He was aware of, and perhaps shared, the local hostility toward some of these families, who, according to their neighbors, had "mean, despicable natures" and "intermarried to keep their riches and property in the family" (Vessie, 1932:565). He knew of twentieth century folk practices of calling people with chorea witches, practices which made more plausible the claim of a seventeenth century connection. "In the

* During the famous Salem witchcraft hysteria of 1692, the young girls who made accusations against the "witches" were the ones who suffered from "fits" and contortions, not the women accused. Indeed, there is no evidence linking any of the accused Salem women to Huntington's disease, despite many claims to that effect.

same community, where these colonial witches lived," he wrote, "there still stands in a woodland region a cabin, called many years ago 'the witch house', for in it there was isolated by relatives a choreic woman whose fantastic muscular movements, combined with a fiery antagonism, attracted considerable attention, and provoked persecution" (Vessie, 1932:563). It was as if Vessie took the twentieth century metaphorical use of the term witch and literalized it, projecting it back into the seventeenth century when the term had quite a different meaning.

In the end, Vessie presented no credible evidence, either genealogical or historical, to support his claim that the women of Bures were "the lamentable means of transporting a family disease from England to the colonial states," nor that the men suffered from this disorder. As Hans and Gilmore (1969:12) put it, "the early cases of Huntington's chorea remain unidentified. Vessie's diagnostic criteria of meanness of character and suspicion of witchcraft are thin at best." So far as we know, none of the accused Connecticut witches — whose histories have been extensively studied by scholars — showed behaviors evocative of Huntington's chorea, nor did any of the men or their early descendants. Moreover, Vessie failed to link any twentieth century families with Huntington's to colonial New England ancestors accused of witchcraft. He argued by association rather than from evidence, juxtaposing descriptions of witchcraft and witch hunting with accounts of the alleged ancestors of his present-day patient with Huntington's chorea, thereby implying a connection. In this sense, Vessie used the witchcraft accusation as it was used historically, as a label to denigrate and devalue. He made no effort to ascertain the situation from the accused witches' points of view. As Vessie made explicit in his 1939 paper, the point of the story for him was the need for "rigid sterilization" (Vessie, 1939:596–600).

NARRATIVE APPEAL(S)

It is perhaps not surprising that Vessie's narrative became "a standard reference in the field and [was] cited in most works which allude to historical aspects of the disease" (Hans and Gilmore, 1969:6–7). The history of hereditary chorea had intrigued clinicians ever since George Huntington (1872:320) had speculated that the disease "was an heirloom from generations away back in the dim past," and had probably come to America with the earliest settlers. Vessie's claim to have traced the disease in North America back over 300 years to an English village in the seventeenth century clearly fascinated clinicians and geneticists in the twentieth. In this "true American tragedy," as he called it, Vessie (1932:556) offered a morality tale with a beginning, a middle, and an open ending, located in a specific geographical locale, with a host of colorful villains and victims.

Here was a cautionary tale, a satisfying allegory, in which the sins of the fathers were visited upon the children "even unto the twelfth generation" as John Terry Maltsberger (1961:1) titled his elaboration of Vessie. Vessie had turned a devastating disease into a story that resonated with Biblical prophecies and romantic tragedy. Here, too, was the illusion of an historical beginning, an origin, and an explanation of how it all began — in the United States at least — which is perhaps why the story also appealed to U.S. families afflicted by this disease.

Moreover, for clinicians, Vessie had created a professionally and emotionally satisfying account of progress, contrasting "these spiteful, furious, unyielding natures of witchcraft days" with "modern observations in the clinic" and noting that "obviously, the occultist does not share the less supernatural viewpoint of the neuropathologist, that a selective destruction of certain parts of the brain develops during adult life, and that the anatomic change is due to an inherited incapacity for continued functional activity" (Vessie, 1932:563–564). In the absence of effective medical treatment, Vessie also specified a role for the physician: "to warn all such choreics and their children against propagation" (Vessie, 1932:565; 1939:600). Although eugenics was losing its appeal among many in the medical profession by the 1930s and 1940s because of the successes of environmental approaches to illness and the difficulty of ascertaining hereditary patterns for most conditions, neurology was probably the specialty most open to eugenic appeals. According to historian Allan Rushton (1994:97), "there was general agreement from the earliest days of the specialty that much disease in this realm of medicine was hereditary." In addition, Charles Davenport's 1911 text, *Heredity in Relation to Eugenics*, had been influential among physicians as a source on human heredity and flatly stated in relation to Huntington's chorea that "persons with this dire disease *should not have children*" (Davenport, 1911:102; emphasis in the original).

Moreover, eugenics was still part of most high school and college biology texts, taught in courses throughout the 1930s and 1940s. Those who read Vessie's two papers within the context of earlier eugenic family studies such as the Jukes and the Kallikak family in the United States (Rafter, 1988) and the "pauper pedigrees" in Britain (Mazumdar, 1992) would have recognized similar concerns in all these narratives — the dangers of intermarriage (such as marriage between cousins), the supposed high fertility of the "dysgenic" classes, and the persistence of "degenerative evolution." Thus, Vessie was playing with familiar associations that were unlikely to be questioned. Certainly Huntington's chorea was one of the conditions for which eugenic recommendations continued to have credibility among physicians.

Even after Vessie's genealogical claims had been discredited and eugenics had fallen out of favor, MacDonald Critchley helped keep Vessie's story alive. As the author of several papers on Huntington's chorea and the president, in the 1970s, of the World Federation of Neurology, Critchley embellished his account with each retelling, calling the descendants of the three Bures couples "undesirables and ne'er-do-wells," many of whom were "liable to bear the marks of a grossly psychopathic taint . . ." (Critchley, 1964:214). Presiding over the landmark 1972 Centennial Symposium on Huntington's Chorea, Critchley (1973:15) stated that the earliest American cases of Huntington's chorea were identifiable "by reason of their sociopathic traits and their criminality." Indeed, for Critchley (1973:15), Vessie's three original families were responsible for many cases "of witchcraft, interbreeding, and crime," which identified them "as Huntingtonians." As late as 1984, Critchley was still rehearsing this lurid history, emphasizing the responsibility of Vessie's families for generations of "intermarriage, illegitimacy and incest" (Critchley, 1984:726).

THE SOCIAL IMPACT OF EUGENIC NARRATIVES

Only recently have scholars begun asking what such eugenically informed medical narratives, and indeed what eugenics itself, may have meant to those who were the subjects of eugenic scrutiny. Sociologist Nicole Hahn Rafter (1988) has argued cogently that the eugenic family studies effectively objectified the subjects of their research, erasing their voices, denying their subjectivities and points of view, and turning them into caricatures in their own eugenic fables. Biologist Nancy Gallagher (1999), in her study of eugenics in Vermont, also notes that the eugenic studies of that state may well have made those described in the studies — including a large family with Huntington's disease — more vulnerable to the prejudices of their communities and more subject to unwanted scrutiny and surveillance. In *Building a Better Race: Gender, Sexuality, and Eugenics from the Turn of the Century to the Baby Boom*, historian Wendy Kline (2001) points out that the people targeted by the early twentieth-century eugenicists may sometimes have internalized eugenic ideology, believing that they had no right to marry, to have children, or, I would add, to even exist. "By arguing that only select bodies were fit to reproduce, eugenicists added unprecedented pressure toward cultural uniformity and contributed to the increasing stigmatization of difference," writes Kline (2001:90). "Restricting motherhood to those who fit uniform standards of mental, moral, and physical fitness, eugenicists succeeded in inculcating a new 'reproductive morality'; this standard convinced some Americans who did not meet it that they should not reproduce" (Kline, 2001:90).

Families with Huntington's disease were especially vulnerable to eugenic propaganda, and Vessie's narrative fit into the tradition of eugenic family studies in erasing the voices of sufferers while recording the views of those who condemned them. Even if affected families may have been intrigued by the notion of the persecuted witch as a woman with chorea, Vessie's melodramatic representation of witchcraft could easily deepen a sense of shame and social stigma, strengthening a determination to hide the disease as much as possible. Moreover, Vessie went beyond merely juxtaposing chorea and witches, constructing an aura of menace, dread, and blame that made Huntington's seem something more than just a disease. He framed Huntington's chorea as a sign of danger, a conventional cultural mode of dealing with disability (the disabled figure as monster). As the *Lancet* opined in 1934, "Dr Vessie . . . refers to the 'diabolical evolution' of the disease, and it is not difficult to concede something devilish in a condition so relentless" (*Lancet*, 1933:870).

Vessie's emphasis on female witches and witchcraft also framed women as those most responsible for the disease, even though the medical literature on Huntington's in the 1930s acknowledged that the disease afflicts men and women equally. Figuring women as witches, and linking women and witches with disease, has a long history, which added to the credibility of Vessie's thesis. Critchley's elaboration of Vessie's thesis further emphasized the guilt of women, since Critchley located the origin of Huntington's in North America in the alleged mother of the three men of Bures. He repeatedly castigated "that local light o'love 'Mary H'" whose "sinister charms" had seduced an unwitting male. She was, in his view, "the villainess of the piece and the probable source of the tainted germ-plasm" (Critchley, 1934:579, 1964:215). Critchley continued to support that theory, even after two English researchers, Adrian Caro and Sheila Haines (1975), discredited it in the 1970s. For both Vessie and Critchley, women were the ultimate source and origin, the Ur-Mothers of the disease.

While those at risk of Huntington's did sometimes wish to avoid passing on the disease to future generations, sterilization was not the only option. Access to birth control and legalized abortion were options that placed more choice in the hands of affected families, especially women. By the 1920s even some eugenicists supported access to birth control, while birth control conferences also gave support to eugenics (Gordon, 1974). However, neither birth control nor legalized abortion had any place in the narrative of Percy R. Vessie. By excluding these options, as well as any discussion of the healthy years that a person might live before symptoms began, Vessie presented a vastly foreshortened picture of the possibilities for families with this disease.

CONCLUSION

With his sensationalized account of witches and criminals as the progenitors of Huntington's disease in the United States, Vessie and those who followed him clearly played to the gendered prejudices of his largely male medical readers. The fact that few clinicians questioned Vessie's account suggests that they shared some of his assumptions and beliefs. Indeed, the British geneticist Peter Harper was the first, in 1991, to go beyond the limited genealogical critique of Hans and Gilmore to condemn Vessie's "lurid and exaggerated descriptions" (Harper, 1991:15). Vessie's powerful myth of origins also owed its appeal to its roots in allegory and fantasy. Had Vessie's historical and genealogical claims been more compelling, we might still question his mode of presentation and the language and rhetoric of his argument. Even if his evidence had been more substantial and his language less prejudicial, we might still consider how the narrative form of the story he proposed, a story of the sins of the fathers (and mothers) haunting the children and the children's children, may have shaped his conclusions. Despite its publication in a scientific medical journal, Vessie himself acknowledged that his history was "a true American tragedy," perhaps alerting us to the fictional forms inhabiting this tale. To suggest, as historians Donna Haraway (1989), Hayden White (1978), and others have done, that factual accounts may follow the forms of fiction and myth is not to denigrate their status as science or history. It is rather to argue for the need note the figurative structures within such narratives, to analyze their criteria of inclusion and exclusion — to ask "what stakes, methods, and kinds of authority are involved" (Haraway, 1989:12). It is to recognize that scientific practice, too, is "a kind of story-telling practice—a rule-governed, constrained, historically changing craft of narrating the history of nature" (Haraway, 1989:4–5).

What White (1978) writes about historical discourse also illuminates medical narratives such as those of Vessie and Critchley. As White (1978:106) writes,

> historical discourse should not be regarded as a mirror image of the set of events that it claims simply to describe. On the contrary, the historical discourse should be regarded as a sign system which points in two directions simultaneously: first, toward the set of events it purports to describe and, second, toward the generic story form to which it tacitly likens the set in order to disclose its formal coherence considered as either a structure or a process.

While scientists and clinicians cannot be expected to critique the evidential basis for historical claims such as Vessie's, sensitivity to the figurative and formal aspects of scientific and medical discourses can make us more critical readers of these narratives that have implications for real people's lives. As White emphasizes, "facts do not speak for themselves;" the historian speaks on their behalf "and fashions the fragments of the past into a whole whose integrity is — in its representation — a purely discursive one." While historians, unlike novelists, deal with real events, "the process of fusing events, whether imaginary or real, into a comprehensible totality capable of serving as the object of a presentation is a poetic process." In other words, the historian must assemble the fragments in the same way as a novelist, "to display an ordered world, a cosmos, where only disorder or chaos might appear" (White, 1978:125).

A critical reading of narratives such as Vessie's and Critchley's can alert us to their figurative and metaphorical strategies, which seduce us into accepting their frames of reference and their conclusions while making it difficult to consider alternatives. Vessie's presentation of "a true American tragedy" foreclosed other readings of the story of Huntington's, as well as options other than the need for "rigid sterilization." Purporting to describe a stigmatizing reality, such narratives helped produce this reality, adding to a repertoire of negative images of people with odd movements and behaviors, framing them as harmful and dangerous and essentially "blaming the victim." Figuring the story of Huntington's chorea as an "American tragedy," neither Vessie, Critchley, nor Maltsberger considered the usefulness of such prosaic possibilities as research into therapies for this disease, even though some of the symptoms, such as depression, were targets of therapeutic interest at the time. Nor did they consider the possible benefit of educating families and the public about this disease, though such strategies had been suggested in the 1920s and would be pursued by family activists beginning in the 1970s (Hughes, 1925). Indeed, many of the insights about people with Huntington's that emerged after the 1970s (for instance the recognition that social contacts and physical activity can help alleviate depression and even slow down cognitive decline) depended not upon new technologies or scientific advances but upon a willingness to listen to the voices of affected families. In short, the type of story Vessie wished to tell helped prefigure his conclusions, turning his narrative toward punitive solutions while excluding more democratic and respectful options and alternatives.

While the idea of involuntary sterilization for those with Huntington's disease is fortunately no longer taken seriously today, the argument is sometimes made that those at 50% risk of such a devastating disease have a moral obligation to take the predictive test, which became available in 1986. However, the expectation that people at risk should get tested, and that those who do not wish to know their genetic status or who know they carry the gene, should not have children, continues the eugenic orientation of earlier decades, even if some of us at risk do make this choice. Unpacking the misogynist and eugenic underpinnings of medical narratives such as Vessie's and Critchley's may help us dispel the shame and stigma attached to diseases such as this, enabling us to think more clearly about how to cope with the dilemmas they present. Moreover, reading medical and scientific narratives with a critical eye, not only their unstated assumptions about gender and eugenics, but also their figurative and formal elements, may help us imagine better stories for the future.

REFERENCES

Aronson, S., Witchcraft was not sorcery but a too human disease, *The Providence Journal*, July 22, 7, 2002.
Buchan, W., *Domestic Medicine*, Garland, New York, 1985.
Bushman, R.L., *From Puritan to Yankee: Character and the Social Order in Connecticut, 1690-1765*, Norton, New York, 1967.
Carlson, C., *The Devil in the Shape of a Woman: Witchcraft in Colonial New England*, Norton, New York, 1987.
Caro, A. and Haines, S., The history of Huntington's chorea, *Update*, 7, 91–95, 1975.
Critchley, M., Huntington's chorea and East Anglia, *Journal of State Medicine*, 42, 575–587, 1934.
Critchley, M., Ed., *The Black Hole and Other Essays*, Pittman Medical, London, 1964.
Critchley, M., In *Advances in Neurology: Huntington's Chorea, 1872-1972*, Vol. 1, Barbeau, T.N.C. and Paulsen, G.W., Eds., Raven Press, New York, 1973.
Critchley, M., The history of Huntington's chorea, *Psychological Medicine*, 14, 725–727, 1984.
Davenport, C.B., *Heredity in Relation to Eugenics*, Henry Holt, New York, 1911.
Davenport, C.B., Huntington's chorea in relation to heredity and eugenics, *American Journal of Insanity*, 73, 195–222, 1916.
Demos, J., *Entertaining Satan: Witchcraft and the Culture of Early New England*, Oxford University Press, New York, 1982.
Gallagher, N., *Breeding Better Vermonters: The Eugenics Project in the Mountain State*, University Press of New England, Hanover, NH, 1999.
Gordon, L., *Woman's Body, Woman's Right: Birth Control in America*, Penguin, New York, 1974.
Hans, M.B. and Gilmore, T.H., Huntington's chorea and genealogical credibility, *Journal of Nervous and Mental Diseases*, 148, 5–13, 1969.
Haraway, D., *Primate Visions: Gender, Race and Nature in the World of Modern Science*, Routledge, New York, 1989.
Harper, P., Ed., *Huntington's Disease*, W.B. Saunders, London, 1991.
Hughes, E.M., The social significance of Huntington's chorea, *American Journal of Psychiatry*, 4, 537–574, 1925.
Huntington, G., On chorea, *Medical and Surgical Reporter*, 26, 317–321, 1872.
Kline, W., *Building a Better Race: Gender, Sexuality, and Eugenics from the Turn of the Century to the Baby Boom*, University of California Press, Berkeley, 2001.
Lancet, Witchcraft and Huntington's chorea, Editorial, ii, 869–870, 1933.
Literary Digest, The witchcraft disease, Editorial, 27–28, 1933.
MacDonald, M., *Mystical Bedlam: Madness, Anxiety and Healing in Seventeenth Century England*, Cambridge University Press, Cambridge, 1981.
Maltsberger, J.T., Even unto the twelfth generation: Huntington's chorea, *Journal of the History of Medicine and Allied Sciences*, 16, 1–17, 1961.

Mazumdar, P.M.H., *Eugenics, Human Genetics, and Human Failings: The Eugenics Society, its Sources and its Critics in Britain*, Routledge, London, 1992.

Mead, S., *Ye Historie of Ye Town of Greenwich*, Knickerbocker Press, New York, 1911.

Muncey, E., *'Classical' Huntington's disease families in the United States*, in rpt. Reed, S.C., Ed., Minneapolis, Dight Institute for Human Genetics, [1964], 1913.

Myerson, A., Ayer, J.B., Putnam T.J., Keeler, C.E., and Alexander L., *Eugenical Sterilization: A Reorientation of the Problem*, McMillan, New York, 1936.

Paige, L.R., *History of Cambridge, Massachusetts 1630-1877 with a Genealogical Register*, Boston, 1877. Reprint Heritage Books, Bowie, Maryland, 1986.

Rafter, N.H., Ed., *White Trash: The Eugenic Family Studies, 1877-1919*, Northeastern University Press, Boston, 1988.

Reilly, P. R., *The Surgical Solution: A History of Involuntary Sterilization in the United States*, Johns Hopkins University Press, Baltimore, 1991.

Rushton, A.R., *Genetics and Medicine in the United States, 1800-1922*, Johns Hopkins University Press, Baltimore, 1994.

Thomson, R.G., *Extraordinary Bodies: Figuring Physical Disability in American Culture and Literature*, Columbia University Press, New York, 1997.

Treichler, P., AIDS, gender, and biomedical discourse: current contests for meaning, In *AIDS: The Burden of History*, Fee, E. and Fox, D.M., Eds., University of California Press, Berkeley, 1988, pp. 190–266.

Vessie, P.R., On the transmission of Huntington's chorea for three hundred years: the Bures family group, *Journal of Nervous and Mental Disease*, 76, 553–573,1932.

Vessie, P.R., Hereditary chorea: St. Anthony's dance and witchcraft in colonial Connecticut, *Journal of the Connecticut State Medical Society*, 3, 596–600, 1939.

Vine, B., *The House of Stairs*, Harmony Books, New York, 1988.

Walker, D.P., *Unclean Spirits: Possession and Exorcism in France and England in the Late Sixteenth and Early Seventeenth Centuries*, University of Pennsylvania Press, Philadelphia, 1981.

White, H., *Topics of Discourse: Essays in Cultural Criticism*, Johns Hopkins University Press, Baltimore, 1978.

6

CONSTRUCTING AND DECONSTRUCTING THE "GAY GENE": MEDIA REPORTING OF GENETICS, SEXUAL DIVERSITY, AND "DEVIANCE"

Jenny Kitzinger

CONTENTS

INTRODUCTION

My mother made me a Lesbian.
If I give her the wool will she make me one too? (graffiti)

The graffiti cited above plays with, and subverts, psychological theories about the causes of lesbianism. It refuses to blame mothers for causing their daughters' homosexuality. Indeed, it inverts the assumption that being a lesbian is a problem to be explained and refuses the very premise that justifies enquiries about the etiology of homosexuality. The joke is on the experts who seek to pathologize sexual diversity. This graffiti reflects a long history of lesbian and gay men's engagement with, and resistance to, attempts to account for their "deviance."

Sexual deviance is often presented as more than a psychological malfunction. It is also (or instead) seen as being written on the body itself. Just as criminal tendencies used to be discerned from primitive physical features (see Dingwall et al., this volume), so homosexual inclinations have been detected by careful attention to the physique and physical development of the suspect. Such ideas are not just imposed on "deviants" by a heterosexual majority; gay scientists, social theorists, and writers have actively promoted ideas about the biological origin of (or elements in) homosexuality. For example, Karl Ulrich, the mid-nineteenth century lawyer, amateur scientist, and gay activist, campaigned to reform the laws on sodomy. He argued that gay men had male bodies but female minds because of faulty fetal development (cited in Rose, 1996).* Similarly, in the 1920s, Radclyff Hall's pioneering and, at the time, highly controversial book, *The Well of Loneliness*, appealed for tolerance on the grounds that lesbians were an intermediate sex "flawed in the making." Hall's book was endorsed as "scientifically accurate" by the leading sexologist of the day, Havelock Ellis (Kitzinger, 1987:120).

In the past it was the enlarged clitoris or strong, masculine hands of the invert that betrayed her as a lover of her own sex and the high-pitched voice, wide hips, and limp wrist of the gay man that revealed his true nature. Today, ideas about the role of biology in homosexuality are likely to be explored through more subtle associations between behavioral tendencies and prenatal hormone exposure, brain chemistry, or genetics. Recent studies into the gay body include, for example, research with

* Fundamental ideas about the way in which sex, gender, and sexuality are linked are deeply implicated in such theories. Sexual orientation is usually assumed to be dimorphic: men programmed for attraction to women, and women usually programmed for attraction to men. "Male homosexuals are thus assumed to have female programming, and lesbians to have male programming" (Peterson, 1999:172).

animals such as genetically engineered "gay" fruit flies (Ryner et al., 1996). It also extends to observations of human beings, exploring gay brain structures (LeVay, 1993), differences in finger lengths (Williams et al., 2000), or features such as the "partial masculinization" of the inner ear of lesbians (McFadden and Pasanen, 1998). The most high-profile recent example of research into the biology of homosexuality, however, occurred in 1993 when Hamer and colleagues published a crucial scientific paper asserting an association between an area of the X chromosome (Xq28) and male sexual orientation (Hamer et al., 1993). This paper provoked widespread controversy, strong reactions from the religious and moral right, and a renewed debate within the gay community. In typical subversive style, it was greeted by some gay men beginning to wear T-shirts bearing the legend "Xq28 - thanks for the genes, Mom!"

In this Chapter I analyze how the U.K. press represented the discovery of what came to be dubbed "the gay gene" and explore the validity of criticisms thrown at the media by political analysts on the one hand and scientists on the other. I then use this case study to reflect on wider issues about the relationship between science and the media. This relationship generates a great deal of heat. Scientists routinely complain about media (mis)representation and consequent public misunderstanding of their work. Journalists are blamed for creating scares about scientific experiments, leading to unnecessary public concern and conjuring up alarmist science fiction visions of the future (Kitzinger and Williams, 2005), sensationalizing and over-simplifying scientific discoveries, giving undue attention to mavericks, and introducing inappropriate prejudices (Williams et al., 2003). The media profile of research into human genetics has been a key focus of such criticism and there have been intensive efforts to reform journalists (or train scientists to be better communicators).

The human genome project, for example, involved major international initiatives and huge economic implications. It raised cutting edge ethical issues and provoked public disquiet. It is hardly surprising, therefore, that those investing in, and developing, such scientific innovation engaged in intensive efforts to influence its media profile and to inform, shape, or encourage public understanding and debate. Much of the attack on the media's role, however, is based on misapprehensions about what journalists should, and could, be doing. The focus on journalists getting it wrong can also help evade key questions about how scientists should relate to society.

In the case of the gay gene story, I argue that attacks on the media for promoting genetic determinism and homophobia are largely misdirected. We need a more considered and focused critique of how the dispute played out. My analysis demonstrates that the mass media, in the United Kingdom at least, are no more deterministic or homophobic than science journals and are arguably *better* at locating the research in its social context and discussing its limitations, assumptions, and implications. A totally pessimistic reiteration of standard critiques of the media is not only inaccurate, but it also obscures important points of potential engagement in this debate. It does not take into account the achievements of pressure groups (in this case predominantly gay male activists) and, at the same time, fails to attend to the more subtle problems of contemporary reporting, including the exclusion of feminist voices and the promotion of a liberal erasure of gay identity under a celebration of human sexual plasticity. Above all, ritually scapegoating the media is problematic because it misassigns blame, letting scientists off the hook and preserving the image of science as a value-free activity that is then misrepresented and distorted by journalists.

A BRIEF HISTORY AND POLITICS OF BIOLOGICAL THEORIES ABOUT HOMOSEXUALITY

Concepts of nature are always used in the performance of culture. Any scientific investigation into genetic diversity raises issues about which bodies, characteristics, or practices attract scientific attention and under what circumstances these are explored. Such investigations also beg questions about how diversity is categorized and the way in which findings are socially interpreted. These questions are particularly pertinent for biological theories about homosexuality. For some critics the very notion of searching for a cause of homosexuality is suspect. Sedgwick, for example, argues, "There currently exists no framework in which to ask about the origins or development of individual gay identity that is not already structured by an implicit, trans-individual Western project or fantasy of eradicating that identity" (cited in Peterson, 1999:165; see also Kaplan, 2000:104). Whatever the motivation behind the search for a cause, the *consequences* of this search have profound social implications.

For a start, notions about causation are deeply implicated in personal biographies for many lesbians and gay men in contemporary Western societies. "Am I really gay?" and "What made me this way?" are questions that face the teenage boy falling in love with a male classmate or the fifty-year-old married woman suddenly lusting after her best friend. "Where did I go wrong?" is an equivalent question confronting parents whose children "come out." These questions can be inverted ("What makes you straight?"). They can also be refused, resisted, or subverted. However, they cannot be completely avoided.

Beyond the private and interpersonal realm, ideas about the causes of homosexuality also have extensive political resonance, carrying with them implications about what should be done about "the problem." Different ideas about the causes of homosexuality are associated with different ideas for control, punishment, or treatment — whether that involves exorcism, incarceration, electric shock treatment, religious counseling, legal "protection," therapy, or tolerance.

Gay men and lesbians thus engage with questions about the causes of their deviance as individuals, as citizens, as friends, as parents, as sons and daughters, and collectively within self-defined communities. Often such engagement is self-consciously political. Indeed, debates about the nature of both sexual desire and sexual identity are at the heart of much feminist theorizing (Jackson, 1981; Kitzinger, 1987). Such issues are also hotly debated within the shifting rainbow alliances and oppositions between gay, lesbian, transgendered, bisexual, and queer communities (Mort, 1994; Weeks, 1994). In attempting to locate a genetic element in homosexuality, genetic science thus enters an explosive arena not only of personal biographies, but also in relation to key policy geographies and the political landscape. Lesbians and gay men can be highly sophisticated analysts and operators, initiating and responding in diverse ways to competing theories about our bodies, sexualities, and identities.

Some lesbians and gay men argue in favor of emphasizing biological explanations. This accords, they say, with the scientific evidence and/or resonates with their personal experience of feeling fundamentally different at a very early age. To ignore the embodied self, and the potential role of biology, is to cut ourselves off from understanding the range and complexity of our own experience. In any case, there are good socio-political arguments in favor of promoting biological explanations for homosexuality. Such explanations, they argue, are preferable to the alternative theories that present us as the product of poor parenting (normal development distorted by a distant father or over-powering mother), the victims of corruption (seduced by predatory older men or women), or as self-indulgent sinners choosing to engage in sheer, willful wickedness.

The idea that we are born gay can relieve lesbians and gay men (and their parents) from guilt. It is also used to invalidate psychotherapeutic interventions or any attempts to make people "go straight." If gays are born, not made, then legislation introduced to protect young people from being "misled" can be challenged. There is no need for an unequal age of consent to protect boys who are uncertain about their sexuality from being seduced into the wrong path, nor can the state justify restricting gays and lesbians from working or living with children (hence challenging job discrimination and the long history of lesbians and gay men losing access to their children). Over and above this, some activists hope that locating biological determinants of homosexuality might bestow scientific authority on calls for tolerance and might have specific legal consequences. In the U.K. context, ideas about the genetic basis of homosexuality were used to challenge the notorious Section 28 (U.K. legislation against the "promotion" of homosexuality). In the context of the United States, it was hoped that establishing homosexuality as an immutable characteristic might extend to homosexuals the constitutional protections of the Fourteenth Amendment. For all these reasons then, the notion of a gay gene was welcomed.

However, some lesbians and gay men beg to differ. They argue that such biological explanations of homosexuality are inaccurate, inadequate, or unhelpful and might even collude with homophobia and heterosexism. Biological theories may not accord with their own experiences of sexual transitions and choices (or, indeed, their experiences of gay sex with "straight" people). Such theories may also not be seen as being based on good science.* Alternatively (or additionally), such explanations may be rejected on strategic grounds. Theories of natural differences have never guaranteed equality; if this were the case, then racism and sexism would no longer be a problem, and discrimination against those born with an impairment would be unthinkable (see Shakespeare, this volume). It is thus, they argue, naïve to hope for progressive policy-making on the basis of discovering a genetic basis for homosexuality. On the contrary, the discovery of the gay gene could feed into oppressive responses such as calls for eradication through abortion or genetic preselection. In any case, the notion that lesbians and gays should be tolerated because they can't help it is hardly a progressive position. Rather than assure normal society that we are not a threat, some lesbians and gay men argue for a more confrontational affirmation of our politics and identities. The radical direct-action group, Lesbian Avengers, for example, refuses to be cowed into claiming that lesbians are born, not made. Their T-shirts defiantly declare, "Lesbian Avengers - we recruit."

Those who challenge the search for a biological basis for homosexuality often draw on (and contribute toward) a more sociological, anthropological, feminist, and/or historical analysis. This includes work that explores homosexuality as a potential choice for anyone. A feminist badge popular in the late 1970s, for example, declared "Any woman can be a lesbian," and pamphlets circulated at the time discussed lesbianism as a political choice to advance women's liberation. Several influential theorists have presented an archaeology of knowledge about homosexual acts and their location within discourses about sin, crime, or sickness across different times and cultures (Weeks, 1994; Plummer, 1981). This approach was increasingly

* The research by Hamer and colleagues was challenged by subsequent studies (see Rice et al., 1999). This led to subsequent headlines a few years later such as "Scientists Cast Doubt on 'Gay Gene' Theory" (*Independent,* April 23, 1999), "Scientists Dismiss Gay Gene Theory" (the *Times,* April 23, 1999), and "No Tears for Passing of 'Gay Gene'" (*Observer,* April 25, 1999). However, this analysis focuses on the original reporting about the Hamer et al. study, and the scientific validity or invalidity of the original work is not the issue that concerns me here.

explored during the 1980s in books such as *The Making of the Modern Homosexual* (Plummer, 1981) and *The Social Construction of Lesbianism* (Kitzinger, 1987). The founding father of such approaches is often identified as Foucault, who famously declared that although previously "the sodomite had been a temporary aberration," in the late nineteenth century "the homosexual was now a species" (Foucault, 1978). The point is, as Mary McIntosh argued a decade earlier, that instead of "seeing homosexuality as a condition that has causes," we should examine it "as a social category rather than a medical or psychiatric one" (McIntosh, 1968).

It is against this background of theorizing and activism that the science of sexual orientation needs to be considered. The paper authored by Hamer and his colleagues that linked Xq28 with male homosexuality appeared in 1993, after over 20 years of gay liberation and feminism as well as a decade of AIDS activism. The following section reviews the intense media interest generated by the research. I start by outlining the criticisms of this press coverage and then go on to assess the reporting itself. I examine the evidence for and against claims that the press coverage was deterministic and homophobic.

CRITICISMS OF MEDIA COVERAGE OF THE GAY GENE AND AN ASSESSMENT OF THE EVIDENCE

The media coverage of the Hamer et al. article was subject to intense controversy. Scientific journals accused the media of "ignoring" the science, "dramatizing" the findings, passing off "inference as fact," and indulging in "fantasies" (see, for example, articles in *Nature* and *Nature Genetics* cited in Miller, 1995:275). Broadsheet journalists (those working on the more "high brow" papers) criticized their colleagues on the tabloids (more popular, or "down market," newspapers) for "hysteria" and "over simplification."*
Critics from a more political perspective attacked the heterosexist bias of coverage (e.g., McKellen, 1993).

Evidence for Determinism and Homophobia in the Press

On one level such criticisms are well founded. It is easy enough to compile a selection of headlines that promoted genetic determinism and presented

* Until very recently, British papers were divided into broadsheet newspapers (traditionally seen as the more serious and up-market outlets read predominantly by the middle and upper classes) and tabloids (more popular newspapers, more likely to be read by lower middle and working class people). The U.K. newspaper scene is also divided into left-of-center newspapers such as the *Guardian* (broadsheet) and the *Mirror* (tabloid), and right-of-center papers such as the *Daily Telegraph* (broadsheet) and the *Daily Mail* (tabloid).

overtly antigay attitudes. Some newspapers immediately dubbed Xq28 "the gay gene," talked of children "born to be gay," conjured up visions of selective abortions, and took the opportunity to reiterate offensive stereotypes. Headlines from tabloid newspapers included "Proof of a Poof" (*Sunday Sport,* July 18, 1993), "Mums Pass Gay Gene to Sons Say Doctors: Parents May Demand Abortions after Tests" (*Sun,* July 17, 1993), and, perhaps the most infamous and disturbing headline of all, "Abortion Hope after 'Gay Genes' Finding" (*Daily Mail,* July 16, 1993).

Several newspapers suggested that there would soon be a clear diagnosis for the gay gene, either *in utero* or in adolescence. Indeed, one report seemed to envisage queues of eager school boys lining up outside the school nurse's office. "Scientists are on the brink of a gene test breakthrough for boys to discover whether they are homosexual," declared the *Sunday Express*: "A simple blood test to determine the adult leanings of boys aged 13-14 could be available within two years." The newspaper added, almost regretfully, that "girls will have to wait until scientists identify the gene linked to lesbianism" ("Schoolboys Could Take a Gay Test, *Sunday Express,* July 18, 1993).

Alongside such genetic determinism some viciously antigay statements were published in the letters pages of both the tabloid and the broadsheet press. The *Evening Standard* printed a letter from the Conservative Family Campaign reasserting that, regardless of claims for a genetic basis for homosexuality, it was clearly "a perversion" (*Evening Standard,* August 2, 1993). Another letter (from Nisson Shulman, Coordinator, Centre for Medical Ethics, Jews College; the *Times*, July 27, 1993) reminded readers that homosexuality was "an abomination" according to the Bible, while a third letter (from E.R. Kermode; *Telegraph*, August 18, 1993) equated the moral standing of homosexuality and pedophilia. Most controversially of all, ex-chief rabbi and member of the House of Lords, Lord Jakobovits, wrote to the *Times* describing homosexuality as "a grave departure from the natural norm which we are charged to overcome like any other affliction" (the *Times*, July 17, 1993). Lord Jakobovits argued that the discovery of a gene for homosexuality did not justify it any more than if there were a gene for kleptomania, adultery, or murder. Jackobvits denied advocating abortion but, in a subsequent letter, argued that "the errant gene" should be "removed or repaired" to free the children of their "disability" (the *Times,* July 30, 1993; this position, of course, reaffirms some analysts' assertions that finding a biological basis for homosexuality would not usher in a new era of tolerance.)

Such vehemently antigay views were mainly confined to half a dozen contributions in the letters pages but were given additional publicity because they generated strong reactions from gay activists and were the focus of several further articles. Although most journalists attacked the antigay attitudes of the letter writers, some critics have argued that the very *structure* of the debate was problematic. As Sir Ian McKellen, founder of the gay rights group Stonewall, comments, "The one overriding issue [covered by science correspondents, leading articles, and the letters pages] was whether a mother should or should not have the right to abort her gay foetus. In other words, what a problem we gays cause our parents" (McKellen, 1993).

Such challenges to the media coverage stake out important rhetorical territory. However, they do not address the full range or even the majority of reporting. In addition to such critiques of media homophobia, it is also instructive to look at how journalists sometimes *challenged* genetic determinism and antigay attitudes and what opportunities there were for more nuanced debate.

Evidence Against Determinism and Homophobia in the Press

Looking beyond the selection of offensive headlines reveals a slightly different picture of media coverage than some of the summaries above imply. Headlines are, of course, penned by editors and subeditors and may not reflect the journalists' reports beneath. In fact, the most homophobic headlines routinely cited by critics do not reflect overtly homophobic articles. A headline quoted by some critics as "advocating" a "search and destroy" mission against gay fetuses, for example, actually reads, "Seek Out and Destroy *Fears*" (emphasis added; see Kitzinger, 1995:310). This article opens with the following statement:

> Gay groups and medical ethics experts yesterday demanded urgent laws to prevent parents aborting babies who may be born homosexual. The uproar followed the Doomsday scenario that scientists could soon be able to tell in the womb if a male fetus is likely to be born gay." (*Mirror*, July 17, 1993)

The piece includes quotes from Stonewall and The Campaign for Homosexual Equality and appears under interviews with "two leading gays": Cashman and McKellan. The captions beneath their respective pictures read: "Wounds: prejudice cuts deep, says Mike Cashman" and "Campaign: human rights at issue says Ian McKellen." The content of this article is a far cry from advocating search and destroy. Similarly, the article beneath the *Daily Mail*'s notorious and offensive headline: "Abortion Hope after 'Gay Genes' Finding" (July 16, 1993) is not quite what one might expect. The headline refers to abortion "hope", but the article focuses on the abortion and screening fears raised as concerns by the gay community. Indeed, these concerns were raised by gay activists *before* any interventions from the moral right publicized the screening and *in utero* treatment "solution" and even though no one openly advocated abortion per se.

Most headlines, and the reports that followed, did not adopt a simple deterministic or overtly homophobic stance. Instead, most journalists pointed to the controversial status of the gay gene and adopted a liberal pro-gay approach. Headlines included:

"'Gay Gene' Claims Spark Anger and Dismay" (*Evening Standard,* July 16, 1993)
"Sex Studies are 'Open to Misuse' — Homosexual Fears" (*Telegraph,* August 17, 1993)
"Genes, Gays and a Moral Minefield" (*Daily Mail,* July 17, 1993)
"Who Is to Judge What Is Normal?" (*Telegraph,* July 17, 1993)
"Why We Should Be Glad to Have Gays" (*Evening Standard,* July 24, 1993)
"Three Cheers for Gays" (*Sunday Telegraph,* July 18, 1993)
"My Fear Is Having Straight Children" (*Independent,* July 17, 1993)

Gay men were high-profile contributors to this debate — both as journalists and as guest writers (e.g., Matthew Parris; the *Times,* July 17, 1993 and Ian McKellen; *Guardian*, July 22, 1993). The coverage also included some striking images of gay activists, with prominence given to their messages — even in tabloids with a reputation for antigay reporting. The right-wing *Daily Mail,* for example, carried a picture of demonstrators holding a placard reading "Justice, equality, freedom for lesbians and gay men" (*Daily Mail,* July 17, 1993). The populist newspaper the *Sun* carried a similar image with the prominent banner "Queer is Cool" and an enlarged inset quote from Stonewall: "Don't try to eradicate us" (*Sun,* July 17, 1993). Gay men were also used as key informants by journalists throughout the reporting across a wide range of newspapers. A systematic study by David Miller of all articles about the gay gene in the national U.K. press on the 16th and 17th of July (around the launch of the scientific paper) identified 29 quotes from scientists, 20 from gay activists, 15 from medical ethicists, and only 4 from the moral right (Miller, 1995: 68). The initial reporting thus clearly identified gay men, rather than the moral right, as the appropriate commentators. This reflects, among other things, the strong relations forged between the media and gay organizations in the context of the AIDS crisis (Miller, 1995). Although Miller does not remark on it, his findings also demonstrate another point: the dramatic exclusion of women's voices from the debate. Only 2 out of 68 quotes recorded by Miller are from women, and my own review of the archives shows that feminist perspectives — on abortion, women's choice, and sexuality — are notable by their absence (see also Steinberg, 1999).

In spite of intense attention to the idea that a gay gene test might lead to abortion, there is not a single example of anyone advocating abortion (as opposed to genetic therapy *in utero*) in any article or letter. This is perhaps not surprising given that right-wing politics, which favor discrimination against gay people, also are often antiabortion. Rather than advocating abortion using a gay gene diagnosis, the very notion of a woman's right to choose on this basis was often presented with horror. One columnist, writing for tabloid newspaper the *People,* for example, declared that only a "warped dysfunctional monster" would consider abortion as a way of avoiding "raising a gentle, caring boy who might — only might, mind you — grow up to love another gentle, caring boy" (cited in Miller, 1995:272).

My analysis of the U.K. press reporting thus tallies with Miller's findings that in the U.K. "the gay gene story was predominantly framed by the assumption that the ethical implications were important and that the potential for discriminating against lesbians and gay men was a serious and feared possibility" (Miller, 1995:272; see also Conrad and Markens, 2001:382).

In addition, the press coverage was not as genetically deterministic as might at first appear. Far from assuming that genes map out our future, many journalists highlighted the role of choice and culture in developing sexual identities and explicitly disassociated themselves from genetic determinism. Headlines included, for example:

"Don't Panic: Take Comfort, It's Not All in the Genes" (*Telegraph,* July 17, 1993)
"It's Not in the Genes, It's in the Culture" (*Independent,* July 19, 1993)
"Gene Talk Won't Wash" (*Today,* July 21, 1993)
"The Myth of the Gay Gene" (*Observer,* July 18, 1993)

The themes in such headlines were explored in the text of these and many other articles. Journalists repeatedly included caveats about Xq28 at the very most influencing but certainly not *determining* sexuality. Readers were informed, for example, that the scientists have "not discovered a gene that causes homosexuality" that "no one expects to discover a gene which 'causes' homosexuality", and even that there is "no such thing as a gay gene" (*Independent on Sunday,* July, 17, July 18,1993; *Daily Mail,* July 17, 1993).

Tabloids and broadsheets sometimes mocked the gay gene story. One journalist commented, for example, "Of course when you start to poke around this tale it tends to sag a bit. It turns out that this gene — which they haven't well, actually, found — might only, possibly, sort of, well, predispose somebody to homosexuality. One person might have it and not be gay, another might not have it and be Julian Clary [a famously camp performer]" (*Sunday Times,* July 18, 1993). Humor was also used

THE SUNDAY TELEGRAPH JULY 18 1993

FOCUS: NATURE v NURTURE

Freud's theory of environment as the major influence on sexuality is cha

BORN TO BE GAY?

The discovery of a 'homosexual gene' aroused the worst fears of gay rights groups. Others, writes Robin Hunt, say it is just a recycled idea

'There has been a tendency to look for biological explanations'

'Heterosexual parents will say: Well, we produced a bad egg'

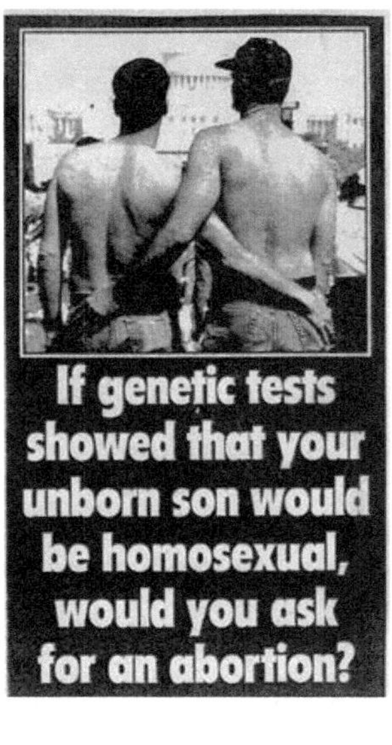

17th July 1993

MUMS PASS GAY GENE TO SONS SAY DOCTORS

Parents may demand abortions after tests

GAY men are born homosexual because of a gene passed on by their mothers, doctors claimed yesterday.

The discovery has smashed the theory that heterosexual men turn gay as they grow older because of outside influences.

By PASCOE WATSON

Family

Society

On the march. . homosexuals at a Gay Pride demo in London last month

'Don't try to eradicate us'

— Gay rights group Stonewall yesterday

Figure 6.1 Although the headlines were often hysterical, much of the media reporting was balanced and sceptical (top: *Sunday Telegraph*, July 18 1993; bottom left: *Today*, July 17 1993; bottom right: *Sun*, July 17 1993).

to challenge the science. The *Independent*, for example, published the following satire:

> Having published their evidence connecting Xq28, an area of the X chromosome, with homosexuality, the team led by Dean Hamer . . . returned to their microscopes and began surveying the contiguous Xq29. Only this time their microscopes were even more powerful than before, and what they saw made them gasp with amazement. What they saw, in perfect minuscule detail, was as follows:
>
> - a complete set of well-used Judy Garland LPs;
> - a complete season of Bette Davis films;
> - Barbara Cartland's complete pink outfit [...]
> - a feather boa [...]
>
> The team realised that what they had stumbled upon was utter dynamite. They had located the genetic determinant of High Camp. (*Independent,* July 19, 1993)

On the few occasions that journalists offered deterministic readings of the gay gene findings, this was sometimes in the service of a distinct agenda. The most extreme deterministic presentation appeared in the *News of the World* in an article that was not focused on the gay gene story at all. Instead, this was a story about smoking. The article was framed in a way typical of this popular and iconoclastic newspaper which often adopts a playful relationship with notions of truth and official expertise, and is at pains to support readers in their own (presumed mainstream and traditional) lifestyle choices. The main point of this article was to reassure readers that it was okay to carry on smoking because the causes of cancer were primarily genetic. The *News of the World* used the example of the the gay gene to prove that genetics has a powerful determining role to play in all aspects of human life. Thus, this article opened by declaring that "American scientists are near to proving that inherited genes via the mother are the cause of homosexuality." This, it stated shows that "what's in our genes sets most of our life patterns" and that lung cancer is "preordained by your genes." Hence, the *News of the World*, declared, the Department of Health is wrong to be "blindly obsessed by the fallacy that smoking's the villain" (*News of the World,* July 18, 1993).

Such simple genetic determinism as was evident in the *News of the World* article cited above was, however, the exception. Indeed, instead of using the gay gene story to promote genetic determinism, several journalists used it in precisely the opposite way. The gay gene research was used as an opportunity to challenge the notion of simple genetic causation. The opening sentence of one article read, "The bogy of genetic determinism needs to be laid to rest. The discovery of a so-called 'gay gene' is as good an opportunity as we'll get to lay it" (*Telegraph,* July 17, 1993).

Several articles and editorials also used the gay gene story as an early opportunity to question the whole enterprise of mapping the human genome and the associated ethical issues. The *Independent on Sunday,* for example, carried an editorial under the headline "The Genetic Tyranny" (July 18, 1993). The gay gene story in this sense offered the opportunity to make human genetic research newsworthy and raise fundamental ethical issues (this opportunity was quite rare back in the early 1990s; see Kitzinger and Reilly, 1997.)

Some newspapers, most clearly the broadsheets, challenged the whole enterprise of searching for a gay gene in the first place. Thus, although few articles questioned the status of the scientific method per se (e.g., statistical significance or sampling method; see Conrad and Markens, 2001), several journalists offered a more fundamental challenge to the science.* They drew on sociological and historical analyses of homosexuality and gay identity (such as those outlined in the introduction to this chapter) in order to question the underlying thesis and politics informing the search for a genetic link.

The morality, validity, utility, and politics of the search for a cause of homosexuality was explicitly questioned in several reports in the U.K. press. One journalist dismissed such knowledge as "useless" and argued that one might as well search for a gene for people who were poor at Morris dancing (*Today,* July 21, 1993). Another cited Stonewall's point of view that the question should not be "Why are we gay?" but "When will we have the right to live our lives without fear of discrimination and persecution?" (*Independent,* July 17, 1993). The *Independent on Sunday* informed readers that "science has had quite a lot to say about homosexuality in the past, most of it rubbish" (*Independent on Sunday,* July 18, 1993). The *Telegraph* referenced historical notions such as that homosexual men have a feminine distribution of body fat (July 17, 1993). The *Guardian* compared the search for a gay gene with efforts to identify a crime gene and its racist implications. This journalist also reminded readers of some of the more ridiculous historic claims for genetics. "In 1919 one respected US biologist

---

* Criticisms of biological research into homosexuality included criticisms about what was defined as homo- or heterosexual activity. For example, in one study the mounting male rats were assumed to be heterosexual even when mounting other male rats (Schuklenk et al., 1998:134). Other criticisms included assumptions about the background rate of homosexuality in the population as a whole (Schuklenk et al., 1998:136), a reliance on small samples, the absence of normal controls, and inappropriate assumptions about what is cause and what is effect (Rose, 1996:58).

published a monograph suggesting that the ability to be a naval officer was an inheritable trait, passed only through the male (because there were no female naval officers)" (*Guardian,* July 17, 1993).

The very notion of the homosexual as a type of being was challenged in some reports. Homosexuality was identified as a social construct. A contributor to the *Sunday Times* commented that the very identity of "the homosexual" was "historically a new idea that even now in many cultures simply doesn't exist" (*Sunday Times,* July 18, 1993). The *Independent* pointed out that "Homosexuality is an abstract noun, a cultural construct with a short historical life, as was famously pointed out by Foucault. Neither the ancient Greeks nor the ancient Romans had a word for homosexuality" (*Independent,* July 19, 1993).* Parts of the media also drew attention to the apparent plasticity of human sexuality. The *Guardian* asserted that human sexuality is a "fluid and creative force" and that "everyone has a capacity" for homosexuality. A *Telegraph* article criticized the gay gene argument because "it polarises homosexuality and heterosexuality rather than seeing them as part of a continuum" (July 18, 1993). A report in the *Observer* illustrated this anecdotally. It opened with the story of a happily married man ("Mr. Nelson") who occasionally liked one-night stands with men. This, according to the journalists, "demonstrates that homosexuality goes far beyond the province of the gay community. After a week of media hysteria about the sources of sexual orientation the experiences of families like the Nelsons should be kept firmly in mind" (*Observer,* July 18, 1993).

CONCLUSION

Reviewing U.K. press reports about the gay gene shows that it is wrong to characterize the coverage of Xq28 as largely naïve and deterministic or as overwhelmingly and uniformly homophobic in any simple sense. The newspapers included some extreme examples of such coverage but were,

* Similar challenges to the authority of "the boffins" is evident in some coverage of the "finger length research." Holliman and colleagues drew attention to a front-page banner advertisement, which linked to an article published further inside that edition which asked the question, "Can your index finger really determine your sexuality?" (*Daily Mail,* March 31, 2000). The article inside answered this question in the headline: "That's handy — the pictures that prove your index finger doesn't determine your sexuality" (*Daily Mail,* March 31, 2000). The paper claimed to put the research to the test so that readers could make their own decisions. This involved a series of pictures of celebrities holding their hands up to the camera. These celebrities had been chosen on the basis of their sexuality, for example: "self-confessed lesbian, tennis champion Martina Navratilova . . . and . . . notorious womaniser Rod Stewart." Almost all the pictures and captions challenged the proposed theory about finger length and sexuality (Holliman et al., 2002).

in general, more cautious in their reporting of Xq28 and less overtly antigay than some critics have implied. I do not wish to be an apologist for the media, but when we generalize about newspapers in ways that are insensitive to differences (e.g., between subeditors and journalists, between letters, pages and editorials, or, indeed, between U.K. and U.S. contexts), we can miss opportunities for intervention. When we reiterate the most pessimistic view of the media, we can fail to acknowledge and build on lobbying successes (e.g., the impact of pressure groups such as Stonewall and Outrage). By focusing on the most obvious and gross examples of homophobia we can also miss more subtle assumptions and exclusions. In this case, for example, it might be worth reflecting on the politics of arguing for a "bisexual continuum" and what this might mean for some sort of postmodern symbolic annihilation of gay identity. It is also worth noting the exclusion from the debate of feminist perspectives on either sexuality or abortion and the way in which pregnant women are framed in the coverage of Xq28 (Epstein, 1999).

The most important point I wish to make is that to scapegoat the media for misrepresenting the gay gene findings is dangerous because it lets science off the hook. When scientists complain about media distortion, they are sometimes refusing to address the social and political implications of their work. When they bemoan media hype, they are failing to acknowledge the public relations activities of their own profession (Rose, 1999). There is no evidence that the media reporting of the Xq28 story was worse than the discussions that took place within scientific circles — and there is some evidence that the media coverage was better. It is true that the media gave Hamer and his colleagues' paper a great deal of coverage, but then the journal *Science* was engaged in extensive prepublication publicity, priming the media that this was a big story, and the lead author made himself widely available for interviews with journalists to promote the findings (Miller, 1995:275).* The press release from Hamer and colleagues did not state that a gay gene had been discovered (Conrad and Markens, 2001:379), but then neither did most of the mass media reporting. However, references to a gene for homosexuality or a gay gene appeared in the title of articles discussing Hamer's work in a range of popular and respectable scientific journals such as *Science* (Pool, 1993), *New Scientist* (Holmes, 1993), and *Scientific American* (Horgan, 1995; see Peterson, 1999). The media focused a great deal on tests and abortions

* For a discussion of factors influencing the coverage, see Conrad and Markens, 2001. They highlight, for example, how reporting of the gay gene in the United Kingdom and the United States was influenced by the prestige of the sources, the "sexiness" of the topic, the context of reporting (e.g., the previous LeVay study), the development of source–journalist relations, and the promotion activities of scientists, as well as different cultural contexts (Conrad and Markens, 2001:39).

(often prompted to do so by gay activists). However, in doing so they drew attention to the social context of such research, and scientists quite explicitly collaborated in this debate. Hamer said he would seek to stop a mythical test (for the gay gene) coming into existence by patenting his discovery, while LeVay (of the "gay brain" fame) added to the debate by saying that he felt bound to support women's choice to abort (on the basis of a mythical test; Rose, 1996:62).

Rather than castigate the media for misrepresenting science or taking it out of context, it is as valid to say that the media reintroduced important social perspectives. As Miller concludes, "The scientific journals' coverage of the debate was significantly narrower than that found in the mass media" (Miller, 1995:269). It was also considerably less reflective. Reporting in the scientific journals, for example, completely failed to explore the distinction between homosexual and gay or acknowledge the historical and social significance of these terms, but the press did, at times, examine this (see Peterson, 1999:171,–175).

Scientists blame the media for hype, yet seeking publicity is an integral part of efforts to promote their research profiles and secure funding (see Kitzinger and Williams, 2005). They accuse the media of rhetorical manipulation but are themselves skilled manipulators (Williams et al., 2003). They blame the media for introducing prejudice yet fail to address the social values informing the scientific enterprise. The gay gene was not invented by the media but was the creation of powerful social forces, scientific efforts, and individual strategies. As Hilary Rose concludes, "The gay gene thesis has been produced not simply by the media misinterpreting science — as scientists commonly claim — but directly through the language and activities of scientists themselves" (Rose, 1996:62). If we wish to problematize the gay gene as a concept, then the media representation is not the main problem. The role of heterosexism and the place of science in society is a more crucial target for challenge. The press can, in some cases, act as a forum for exposing prejudice and pursuing critique and resistance.

ACKNOWLEDGMENTS

The support of the Economic and Social Research Council (ESRC) is gratefully acknowledged. This work was part of the program of the ESRC Research Centre for Economic and Social Aspects of Genomics.

REFERENCES

Blackwood E., Ed., *The Many Faces of Homosexuality: Anthropological Approaches to Homosexual Behavior*, Haworth Press, New York, 1986.

Brookey, R., *Reinventing the Male Homosexual: The Rhetoric and Power of the Gay Gene*, Indiana University Press, Bloomington, 2002.

Cohler, B., *The Mismeasure of Desire: The Science, Theory and Ethics of Sexual Orientation*, Oxford University Press, Oxford, 2000.
Conrad, P. and Markens, S., Constructing the 'gay gene' in the news: optimism and scepticism in the US and British Press, *Health,* 5, 373, 2001.
DeCecco, J.P. and Elia, J.P., Eds., *If You Seduce a Straight Person, Can You Make Them Gay? Issues in Biological Essentialism Versus Constructionism in Gay and Lesbian Identities*, Haworth Press, New York, 1993.
Foucault, M., *The History of Sexuality*, Pantheon, New York, 1978.
Hall, R., *The Well of Loneliness*, Bard Avon Books, New York, 1928.
Hamer, D., Hu, S., Magnuson, V., et al., A linkage between DNA markers on the X chromosome and male sexual orientation, *Science*, July 16, 321, 1993.
Holliman, R., Scanlon, E., and Vidler, E., Reporting contested science: comparing media coverage of genetic explanations for sexuality and intelligence. PCST Conference, December 2002, Parallel Session 25: Communicating Biotechnology and Biomedical Sciences.
Holmes, B., Gay gene test 'inaccurate and immoral', *New Scientist,* 141, 9, 1994.
Horgan, J., Gay genes, revisited, *Scientific American*, 273, 26, 1995.
Jackson, S., *The Social Construction of Female Sexuality.* Women's Resource and Outreach Centre, London, 1981.
Kaplan, J., *The Limits and Lies of Human Genetic Research: Dangers for Social Policy*, Routledge, London, 2000.
Kitzinger, C., *The Social Construction of Lesbianism*, Sage, London, 1987.
Kitzinger, C., The sexual brain by Simon LeVay (book review), *Theory and Psychology*, 5, 309, 1995.
Kitzinger, J. and Reilly, J., The rise and fall of risk reporting, *European Journal of Communication,* 12, 319, 1997.
Kitzinger, J. and Williams, C., Inventing the future: legitimising hope and calming fears in the embryo stem cell debate, *Social Science and Medicine*, 61, 731–740, 2005.
LeVay, S., *The Sexual Brain*, MIT Press, Cambridge, MA, 1993.
McFadden, D. and Pasanen, E., Comparison of the auditory systems of heterosexuals and homosexuals: click-evoked otoacoustic emissions, *Proceedings of the National Academy of Sciences of the United States of America*, 95, 2709, 1998.
McKellen, I., Through a gay viewfinder, *Guardian,* July 22, 1993.
McIntosh, M., The homosexual role, *Social Problems*, 16, 182–191, 1968.
Michaels, S.M., Queer Counts: The Sociological Construction of Homosexuality via Survey Research, Ph.D. thesis, University of Chicago, Chicago, 1997.
Miller, D., Introducing the gay gene: media and scientific representations, *Public Understandings of Science,* 4, 269, 1995.
Mort, F., Essentialism revisiting? Identity politics and the late twentieth-century discourses of homosexuality, In Weeks, J., Ed., *The Lesser Evil and the Greater Good: The Theory and Politics of Sexual Diversity*, Rivers Oram Press, London, 1994.
Peterson, A., The portrayal of research into genetic-based differences of sex and sexual orientation: a study of 'popular' science journals, 1980-1997, *Journal of Communication Inquiry*, 23, 163, 1999.
Pool, R., Evidence for homosexuality gene, *Science*, 261, 291–292, 1993.
Plummer, K., Ed., *The Making of the Modern Homosexual*, London, Hutchinson, 1981.
Rice, G., Anderson, C., Risch, N., and Ebers, G., Male homosexuality absence of linkage to microsatellite markers at Xq28, *Science* 23, 665, 1999.

Rose, H., Gay brains, gay genes and feminist science theory, In *Sexual Cultures: Communities, Values and Intimacy*, Weeks, J. and Holland, J. Eds., St. Martin's Press, 1996, pp. 53–74.

Rose, S., War of the genes: a genetic code that makes you aggressive, schizophrenic, or even straight or gay? *Guardian,* May 8, 1999.

Ryner, L., Goodwin, S., Castrillob, D., et al., Control of male sexual behaviour and sexual orientation in Drosophila by the fruitless gene, *Cell,* 87, 1079, 1996.

Schuklenk, V., Stein, E., Kerin, J., et al., The ethics of genetic research on sexual orientation, *Reproductive Health Matters,* 6, 134, 1998.

Stein, E., Ed., *Forms of Desire: Sexual Orientation and the Social Constructionist Controversy,* Routledge, London, 1992.

Steinberg, D., Pedagogic panic or deconstructive dilemma: gay genes in the popular press, In *A Dangerous Knowing: Sexuality, Pedagogy and Popular Culture,* Epstein, D. and Sears, J., Eds., Cassell, London, 1999.

Weeks, J., Ed., *The Lesser Evil and the Greater Good: The Theory and Politics of Sexual Diversity,* Rivers Oram Press, London, 1994.

Weeks J., *Making Sexual History,* Polity Press, Cambridge, 2000.

Williams, C., Kitzinger J., and Henderson, L., Envisaging the embryo in stem cell research: rhetorical strategies and media reporting of the ethical debates, *Sociology of Health and Illness,* 25, 793, 2003.

Williams, T.J., Pepitone, M., Christensen, S., et al., Finger-length ratios and sexual orientation. *Nature,* 404, 455, 2000.

Wilson, E., *Neural Geographies: Feminism and the Microstructure of Cognition,* Routledge, London, 1998.

7

THE DILEMMA OF PREDICTABLE DISABLEMENT: A CHALLENGE FOR FAMILIES AND SOCIETY

Tom Shakespeare

CONTENTS

INTRODUCTION

Since the early 1970s, disabled people have challenged the traditional individual medical model of disability. Instead of conceptualizing disability as a personal problem, they have sought to relocate the problem to the social level. A rights perspective focuses on barrier removal and anti-discrimination legislation rather than on cures and prevention. However, a rise in genetic technologies and practices may newly challenge the ability of disabled peoples to control their futures. Is the social approach compatible with the increasing development and use of prenatal diagnosis and preimplantation diagnosis? Will the rise in use of these genetic diagnostic technologies increase discrimination against disabled people in society, as well as reducing the numbers of disabled people? Will individuals be blamed if they fail to take advantage of technological interventions? In what circumstances is it preferable not to disclose genetic information? What are the appropriate responses to the problem of disablement?

This chapter explores the ways in which genetic research into impairment and illness may impact on the lives of disabled people and their families and ultimately change the ways in which disability is thought of and dealt with in society. Recent years have seen considerable media and political hype surrounding research and practice in genetics and reproductive technology. For example, "gene for . . ." discoveries are regularly covered in the press, with the implication that the new knowledge will improve human health in the near future (Nelkin and Lindee, 1995). The completion of the working-draft human genome in 2001, followed by the completion of the Human Genome Project in 2003, ahead of schedule and at a total cost of approximately $3 billion, became front page news and was heralded as a milestone for life on earth, not just for biological research (Ridley, 2000). Also in 2003, the 50th anniversary of the publication by Crick-Watson of the structure of DNA was celebrated as a triumph of British science and evidence of how fast genetics is moving. In this arena, scientists and politicians are privileged commentators, and genetics is widely celebrated as a panacea for disease and disability.

However, the voices of people directly affected — disabled people and their families, who may have more mixed and nuanced views of the impact of genetics on their lives — are rarely heard. Of course, these views cannot be represented simply. Disabled people and people affected by different conditions have many different views. For some groups of disabled people — particularly those affected by degenerative conditions — there is hope that genetic research will lead to therapies that might cure disease. In the disability rights community, however, there is considerable anxiety about the use of genetic diagnoses and selective termination technologies to prevent disabled people from being born. Interestingly, claims for "eugenic abuse and "miracle cures" are equally exaggerated and should be evaluated carefully (Shakespeare, 1999), yet behind the fear and the hope are important questions about the implications of diagnosis for prospective parents and for disabled people who are born in the future.

An immediate problem is the absence of empirical work on the impact of genetic research and genetic services on either policies for or attitudes toward disabled people. Longitudinal studies of the ways in which attitudes toward disability and genetic conditions may be changing with the advent of genetics are needed to contribute to bioethics debates and policies. More quantitative and qualitative studies of the understanding and attitudes of different groups of disabled people are needed to explore their views and experiences. In 1999 the Royal Association for Disability and Rehabilitation (RADAR) conducted a survey of the views of disabled people and their close relatives (n = 452). The majority of respondents (53%) felt a mixture of hopes and fears about genetics, although more felt hopeful (22%) than were worried (15%) or horrified (8%). There was considerable support for research that might lead to treatments (73%). A modest majority (55%) thought that abortion should be freely available, and only 9% thought that abortion should be legal on social grounds but not on the grounds of fetal impairment.

I carried out a very small survey of members of the U.K. Restricted Growth Association in 2000 (n = 74) and asked what effect they thought knowing the genetic cause of achondroplasia and related conditions would have.

	Positive Effect	*Negative Effect*	*Not Sure*
People with restricted growth conditions	20.6%	26.5%	32.4%
Nondisabled relatives and friends	40%	7.5%	50%

Two points here seem striking. First, people directly affected were much more likely to think that genetic research would have negative outcomes for people with restricted growth. Second, both disabled and nondisabled respondents reported that they were uncertain about the impact of this genetic discovery.

NAMING AND KNOWING DISABILITY

The development of new powers of genetic diagnosis may have important implications for the process of diagnosing and understanding disability and disease. Even before the social and ethical consequences of diagnosis and treatment are considered, simply creating genetic claims about disability and difference has an impact on culture and society. Three elements of this process have particular relevance to this chapter.

Medicalization of Disability

The disability movement emerged in Britain and the United States during the 1970s but is now a worldwide phenomenon (Driedger, 1989). Disabled people have challenged social exclusion and disadvantage and developed a political identity as disabled people. A major part of this mobilization has been about redefining disability as a social construction rather than an essential biological difference. For example, in U.S. disability politics, the minority group model of disability (Hahn, 1988) argued that the disadvantage of a person with disability "stems from the failure of a structured social environment to adjust to the needs and aspirations of citizens with disabilities rather than from the inability of a disabled individual to adapt to the demands of society" (Hahn, 1988:128). In Britain, this redefinition resulted in the rejection of the World Health Organization's International Classification of Impairments, Disabilities and Handicaps (ICIDH) and the term *people with disabilities*, in favor of what became known as "the social model of disability" and the term *disabled people*. The social model defines disability as a relationship between people with impairment and a society that disables them. This disablement takes the form of social and environmental barriers, discrimination, prejudice, and other structural problems, rather than deficits of body or brain. While the British social model is a stronger version of the social constructionist approach to disability than social definitions developed by disabled people in other countries, it influenced the position adopted by Disabled Peoples International, who defined disability in terms of social barriers, not biological differences (Bickenbach et al., 1999; Driedger, 1989).

Disability activists throughout the world have argued against what they label "the medical model of disability." The ICIDH definitions provide a somewhat narrow interpretation of the medical model, since these make impairment (bodily deficits or abnormalities) and disability (functional incapacities) the conceptual basis of disablement and do not have room for social environments to play a causal role. From a wider perspective, the medical model refers to the tendency to see doctors and related professionals as the appropriate authorities on disability and the idea that rehabilitation, therapy, and cure are the appropriate responses to the disability problem. Culturally, the medical model is the label for a family of stereotypes and attitudes in which disability is a medical tragedy and disabled people are invalid, incapable, and "othered." The social model of disability, in contrast, has been the basis of the positive political identity of many disabled people and has mandated a different approach to disability based on the removal of social barriers and the passing of antidiscrimination legislation.

In this context, increasing tendencies first to see social issues in medical terms and then to see medical issues in genetic terms risk both redefining disability as an individual pathology again and reducing the complexity of disability to the stark simplicity of DNA mutations. Thus, dwarfism becomes an FGFR3 mutation rather than a complex social and cultural experience in which the role of disabling barriers and attitudes creates problems for people with restricted growth. Disability becomes a medical problem, and genetics becomes the high-tech solution, either through prevention or cure.

The way genetics redefines disability, undermining the political identity of all disabled people, as well as the perceived threat of eugenics (Shakespeare, 1999), perhaps explains some of the vehemence of disability rights groups' objections to the new biology. The views of an activist like David Colley, whom I recorded for a radio interview, express this powerfully:

> Disabled people have everything to fear from the new genetics. At worst it's our very existence, that we'll be eliminated simply as genetic spelling mistakes. But even at best we will be reinforced as biological abnormalities, as defective human beings. (Shakespeare, 2003:10)

Two further points should be noted. First, by breaking the definitional link between impairment (bodily difference) and disability (social restriction), social modelers have created a logical problem that makes it harder to challenge genetic science. In other words, by saying that disability is not caused by biology, disability activists risk leaving biology to geneticists and others who wish to eliminate biological differences. Moreover, by generating a powerful dichotomy between society (disability) and biology (impairment), the social model makes it harder to contest the nature or cultural meaning of impairment. Social modelers have fallen into the same trap as the feminists who polarized sex and gender (Tremain, 2002). Second, a second wave of disability-studies perspectives have challenged the social model, in particular the exclusion of impairment (Crow, 1996; Shakespeare and Watson, 2001). A more nuanced account of disability, which combines the individual and structural factors, would overcome the logical problems (Williams, 1999). However, the imperative to challenge the one-sided biologization of disability remains.

Geneticization

The renewed medicalization of disability is part of a broader problem, which is about how genetics impacts our understandings of difference and disadvantage in general terms. Abby Lippman (1992) coined the term *geneticization* to explain a process that she believes is a problem for society as a whole. Complex social and economic problems may be reduced to genetic explanations and responses, because individual genetic explanations are simple, compelling, and perhaps even comforting. Moreover, if commercial interests see business potential in genetic diagnosis, pharmacogenetics, and gene therapy, this acts as an incentive to promote the biological basis of socially stigmatized diseases or behaviors. For example, new drugs are now offered to manipulate a range of personality traits, including melancholy and shyness. Similarly Erik Parens has argued that:

> Some new means that work on our bodies instead of our environments may incline us to ignore the complex social roots of the suffering of individuals. The easier it is to change our bodies to relieve our suffering, the less inclined we may be to try to change the complex social conditions that produce that suffering. (Parens, 1998:53)

Education, counseling, social change, and barrier removal all represent potential costs for welfare states, rather than profits for pharmaceutical companies.

However, there may be a counter argument that in some cases to find organic and even genetic causes of conditions may be welcomed by people experiencing them. For example, when Dean Hamer claimed to have found a marker associated with homosexuality at Xq28, some gay groups and individuals in the United States welcomed the development, believing that it demonstrated that sexuality was not a lifestyle choice but an innate property of individuals (see Kitzinger, this volume), perhaps even one resulting from divine creation (Burr, 1997). Similarly, if behaviors such as alcoholism and other addictions are redefined as disabilities, then some blame may be removed from those succumbing to these problems. In the case of autism, the move from blaming psychodynamic factors arising from poor parenting toward blaming organic brain dysfunction, possibly of genetic origin, has been very liberating for many families, who no longer feel that they have failed their children and may even feel that there is hope of a cure in the long term. Those with uncertain and contested diagnoses — such as dyslexia or chronic fatigue syndrome — will similarly often welcome the medicalization of their conditions, both because it proves that their experience is real, and also because medical explanations raise the (perhaps unrealistic) hope of medical cures.

Genes as Destiny

This discussion has highlighted two important dimensions in our understandings of disability and disease: biological–social and individual–structural. A third dimension relates to the extent to which conditions are regarded as determining and unchangeable or dynamic and potentially controllable. Genetic explanations have the power to redefine conditions and differences as fixed and unchanging. This arises out of the cultural representations of genetics as being about innate biological blueprints or codes. The gene seems to have become the modern version of the soul: immortal and essential. The notion of "one gene, one disease" and possibly the simple patterns of Mendelian inheritance have passed into popular consciousness via the media and secondary school biology, whereas the complexity of gene–gene and gene–environment interaction is harder to explain and to understand (Richards, 1996). The deterministic version of genetics is sometimes fuelled by simplistic media coverage, but also by the extravagant pronouncements of scientists such as James Watson, who many years ago pronounced that "in the past, men thought their fate was written in the stars. Now we know, in large part, it is written in the genes"... or words to that effect.

This sort of genetic fatalism has implications for those who are diagnosed with late-onset disorders or a susceptibility to later disease. Research into the psychology of genetics by Theresa Marteau and her team has shown that patients' responses to a genetic diagnosis are different from their responses to other forms of medical diagnoses (Marteau and Senior, 1997). For example, when a screening test for inherited risk of heart disease was seen as detecting a genetic problem, the condition was perceived as less controllable, but when the test was perceived as detecting raised cholesterol, patients perceived the condition as caused by diet and therefore possible to control.

Contemporary understandings of genetics are more dynamic and subtle than traditional accounts. The revelation that there are perhaps fewer than 30,000 genes in the human genome has sometimes been taken as evidence that genes cannot explain the complexity of human behavior and action. However, the real significance is to highlight the way that gene regulation and gene expression are central to understanding genetics. The switching on and off of genes, via promoter genes, explains important biological differences and is evidence of the continuing importance of environments and choices and the malleability of effects (Ridley, 2004). Moreover, just because a phenomenon is social, not genetic, in origin does not mean that it is easy to change or engineer. Social inequality has proved very persistent, despite welfare states and reforming strategies. Clearly, both social and biological interventions will continue to be necessary in a range of conditions and problems. Neither environmental nor genetic determinism has proven to be helpful.

PRENATAL RESPONSIBILITY AND PARENTAL ANXIETY

The reference to those who might welcome the medicalization of their experience points to a broader question about the ways in which the redefinition of difference and disability discussed earlier has implications for blame and responsibility. These issues are particularly acute in the prenatal situation. The development of more screening and diagnostic tests means that prospective parents have the responsibility of deciding whether to use tests to discover characteristics of their fetus and, if something is discovered, the difficulty of having to decide what to do. This adds to the stress and anxiety of pregnancy for everyone, not just the tiny proportion for whom a genetic condition is diagnosed.

Going back to my survey of the U.K. restricted growth association, I asked parents of people with restricted growth conditions whether they would have had a test in pregnancy, had one been offered:*

	Would Want a Test	*Would Not Want a Test*	*Not Sure*
Parents and friends	32.5%	60%	7.5%

A typical comment of one parent, asked whether she would have had a test, was, "Not now because we had him and we loved him immediately, but had we have known via a test I honestly don't know whether we would have considered termination. All I can say is I'm glad we didn't know, otherwise who knows?" Counter-factual questions are notoriously unreliable, but such comments show that parents often come to terms with the initially unwelcome diagnosis of this genetic condition. Families adapt positively to many unforeseen genetic differences, including more serious conditions such as Down's syndrome (Hall et al., 2000). Human beings show resilience and adaptation in many different and difficult circumstances.

Of course, it needs to be pointed out that in many cases the predictions made by technologies such as ultrasound are uncertain. *Soft signs* may indicate a serious abnormality or they may be a passing problem or an artifact of the scanning technique. Raised levels of a hormone in the blood may indicate Down's syndrome or may be a false positive. A chromosomal abnormality may have no phenotypic effects or may have varying phenotypic effects. In all of these situations, anxiety is inevitably raised, making pregnancy more stressful. However, not only are there no clear answers as to what parents should do or how they determine what is best for their potential child, but often these concerns are raised when there is a high likelihood that the child would be perfectly healthy and unaffected.

Even where there is a clear diagnosis, the abiding moral and social dilemmas are about what action the prospective parent(s) should take

* The genetic cause of achondroplasia was only recently discovered, and antenatal screening is not currently available for restricted growth conditions.

as a result of the information. Does responsibility for a potential child mean in some cases deciding not to have that child? What is a good life, and what degree of impairment makes life "not worth living"? When does life begin? The questions arising from a diagnosis of fetal abnormality are both empirical and normative. Clinicians are meant to adopt an ethic of nondirectiveness. Traditional scriptures and mores offer little guidance on virtuous behavior in these modern circumstances, and there is no simple answer as to what the parents should do. Whether they continue pregnancy or terminate pregnancy, grief will be associated with their decision, and it is one that they will have to live with for the rest of their lives. As Rayna Rapp argues, in the context of amniocentesis, "this technology turns every user into a moral philosopher, as she engages her fears and fantasies of the limits of mothering a foetus with a disability" (2000:128).

There are no right answers to the dilemma facing prospective parents. In most cases, the decision must depend on the circumstances, the characteristics, and the moral values of the prospective parents. Ideally, medical advisors would give full and balanced information and support prospective parents to make the decision that is right for them, in the context of a society that is welcoming of disabled people and supportive of disabled people's rights. When, in practice, medical personnel impose their own values and expectations, or when society does not welcome disabled children, the situation becomes reminiscent of the eugenic abuses of the past. Because disabled people's voices are absent from these encounters, there is little to balance popular prejudice and medicalized and pathological accounts of disability. Disabled people's lives are complex and variable, and social factors are just as important as medical ones: hearing accounts of life with genetic and developmental impairments is vital, in order that prospective parents can understand the implications of the diagnosis they have received. For this reason, I have led a Wellcome Trust–funded project to provide a Web site, the Prenatal Screening Web Resource (ANSWER)*, to supply life stories of individuals and families affected by conditions for which screening is offered.

Aside from the immediate context of reproductive decision-making, it could be argued that one of the impacts of genetics has been to increase personal responsibility for having disabled children and to pave the way for blame being accorded to parents who exercise their choice not to use these screening technologies or not to terminate pregnancies. This way of thinking was graphically illustrated by Professor Bob Edwards, one of the two pioneers who enabled the birth of the first test tube baby in 1988, who said, "In the future, it will be a sin to have a disabled baby" (Rogers, 1999).

* ANSWER can be found at www.antenataltesting.info

Bruce Jennings suggests that prenatal genetic testing shapes choice by making everything into a choice. Furthermore, he argues,

> There is a subtle cultural difference between, on the one hand, the kind of sympathy we give when someone receives sudden bad news that could not have been known in advance, and, on the other hand, the kind of sympathy we give when someone finds out something awful that could have been known before and could have been altered, albeit at a psychological and moral cost. (Jennings, 2000:134)

Theresa Marteau and Harriet Drake's research on attribution for disability, suggests that people are inclined to blame parents who have children affected by Down's syndrome:

> ...both health professionals and lay groups make judgements about women's roles in the birth of children with disabilities. Women who decline the offer of testing are seen as having more control over this outcome, and are attributed more blame for it, than are women who have not been offered tests and also give birth to a child with Downs syndrome. (Marteau and Drake, 1995:1130)

Of course, an increasing number of women are now offered tests, and most people know that there are tests available, so this response is likely to spread. Not just conditions such as Down's syndrome but many congenital conditions have moved from being unfortunate examples of bad luck to blameworthy failures of surveillance and control. I have heard anecdotes of people facing negative or disbelieving responses from family or friends because it is believed that Down's is an outcome that can and should be avoided. What happens to the false negative diagnosis here? What happens to the rights of parents to make up their own minds about ethical issues such as termination of pregnancy or quality control of embryos? Even if information and professional support is scrupulously fair and balanced, the choice that each prospective parent makes is made harder and more predetermined, because the majority of people decide to terminate affected pregnancies (Beck-Gernsteim, 1990).

These pressures do not just derive from the negative views of others or the pressure to emulate the decisions of the majority. We also have to consider the social and economic costs that individuals may have to bear if they decide not to terminate pregnancies. In the future, having a disabled child might come to be seen as *elective disability*, and welfare or insurance systems may refuse to meet the additional costs of disability. This is again part of the individualization of disability that may be the consequence of genetics. Previously, disability was bad luck or "karma" or part of "God's plan." Society compensated individuals for misfortune and took the responsibility to look after its weakest members. Now congenital disability may become the fault of an individual who has to bear the costs because the rest of the community doesn't want to have to pay. Again, Dorothy Nelkin reports that in the United States some local communities have decided that learning difficulties are a medical matter, not an educational matter, so their taxes should not be devoted to the special educational needs of people with learning difficulties, but can be concentrated on ensuring that their nondisabled children get better educational provision.

There is a general point that the development of the concept of *elective disability* makes the reproductive choices of everyone less free, but there is also a specific point that relates to the reproductive choices of disabled people and perhaps also of those minority ethnic communities who have high rates of particular disorders. In the first part of the twentieth century, eugenic programs concentrated on which people should be allowed to reproduce. Postwar genetics has concentrated instead on which babies are allowed to proceed to term. Moving toward greater responsibility and blame for reproductive decisions might return us to a world in which certain people are blamed for having particular children because they now have the choice not to have children affected by particular conditions. Many disabled people fear that this might lead to a eugenic backlash against disabled people, who are regarded as irresponsible if they reproduce and knowingly bring a disabled child into the world. For example, the desire of a deaf couple to find a sperm donor who was also deaf, to maximize their chances of having a deaf child, was widely condemned in the United States and British media (Spriggs 2002). When I was a student at King's College, Cambridge, one of the fellows remarked, on seeing me walk through the college with my baby daughter in her buggy, that it was irresponsible of me to have had a child, knowing it would have a 50% chance of having achondroplasia. Several of my friends voiced similar sentiments. Lots of disabled parents have similar stories to tell.

EFFECTS ON PROVISIONS FOR DISABLED PEOPLE

Prenatal diagnosis and reproductive decision-making are not the only areas of concern. Disabled people will always exist, because not everyone will adopt screening and because screening is fallible and limited. What impact might genetics have on disabled people themselves, on disability services, and on attitudes to disability? Here, empirical evidence is even more scarce and discussion necessarily becomes more speculative. However, it would be foolish not to explore the possible negative effects of genetic research.

Still, it is important to acknowledge the possible benefits of genetics, which are also hypothetical: diagnosis of predisposition to disease might enable people to avoid behaviors that might contribute to ill health; and, in the long term, the aim of research is to generate therapies or pharmaceuticals that can cure or alleviate conditions.

In theory, if genetics succeeds in reducing the numbers of disabled people in society, this might lead to a general reduction in facilities and a slowdown of investment in accessible transport and housing. However, this fear relies on ignorance of the disability statistics. While approximately 1% of births are affected by congenital abnormality, at least 12% of the population are disabled, largely owing to the effects of ageing, accident, disease, and so forth. The majority of disabled people are old, and the global population is rapidly ageing, generating increased morbidity. Much disability has neither a genetic cause nor a genetic solution. Gene therapies and stem cell research are at an early stage, and optimism may not translate into clinical treatments. Treatments will take decades to materialize and may remain inaccessible to most of the world's 450 million disabled people because of the cost. Therefore, any promise by its more enthusiastic advocates that genetic screening will reduce costs or have a major impact on the numbers of disabled people in the population may be misplaced.

The advent of genetic screening might have an effect on research into medical and assistive technologies to help people affected by particular impairments. For example, now that the numbers of people affected by neural tube defects have fallen so steeply in many Western societies, one might hypothesize that expertise and facilities for medical and social personnel dealing with the effects of spina bifida might have been reduced as a consequence. If a medical student was looking for a career specialty, arguably he or she would be foolish to concentrate on an area for which there will be decreasing demand, such as medical management of spina bifida.

There is anecdotal evidence that the life expectancy of people affected by cystic fibrosis may have actually decreased in North America in recent years. If this is the case, it may be a result of less research and expertise being devoted to the care of people affected by a condition that can be prevented by genetic screening. Earlier, it was suggested that genetic screening programs are likely to remain incomplete and fallible, and there will be continued noncompliance. The sad fact may be that, in the cases of certain impairments, people born with these conditions in the future might face worse services than they can currently expect, unless we take steps to address these problems.

Finally, going back to the discussion of our changing views of disability in the postgenomic age, it could be hypothesized that these increasingly rare individuals with conditions may be more isolated in the future than they are now because the public will be less familiar with their differences and there will be fewer people with whom to band together to form self-help organizations or to campaign for greater awareness and better provisions. Of course, this is not an argument for not trying to reduce the incidence of serious problems. If a self-help group for victims of land mines disappeared because land mines had been subject to an international ban, then it would be a cause for celebration. In the case of genetic conditions, there will always be people affected, and we need to pause and realize that their social experience may change as a result of their increasing rarity in society.

FORETELLING THE FUTURE

Moving further away from issues of prediction before birth, there are a whole range of ethical, social, and psychological issues relevant to people who are born healthy but are diagnosed as having late-onset genetic conditions such as Huntington's disease or having increased susceptibility to conditions such as breast or colon cancer, Alzheimer's, and so on. Here there is an irony. Genetics may enable societies to avoid the birth of disabled people and hence have an impact on the numbers of people with impairment in society (albeit not so significant an impact as some have claimed if earlier arguments turn out to be accurate). However, genetic services have the power to turn healthy people into disabled people, by virtue of diagnosing future illness or risk of illness in the currently asymptomatic. As soon as a genetic test has returned from the laboratory, a person's identity, future plans, and psychological well-being may be completely altered. Suddenly, the population of disabled people increases many times over, as we consider all those who have a higher risk of Alzheimer's, diabetes, cancer, and so forth.

The disability movement defines disability in terms of discrimination and prejudice, not simply impairment or disease. People with presymptomatic genetic conditions, or those who are at risk of late-onset conditions, may experience stigma and discrimination from employers, insurance companies, and others. Both because they are "disabled by society" and because they may develop a changed sense of identity, people who are diagnosed with late-onset conditions may find themselves having common cause with disabled people. Exploring the problems of discrimination in insurance and elsewhere for the new group of people with what could be called hidden or latent disabilities is beyond the scope of this chapter, but it might be worth considering how predictive genetics may change the identity and self-image of people who formerly thought of themselves as healthy and nondisabled. A wealth of psychological research with people diagnosed with Huntington's Disease (Cox and

McKellin, 1999) or breast and ovarian cancer (Hallowell, 1999) explores the complex and difficult effects that such prediction or risk calculation has on individuals and families. No wonder predictive genetic information has been referred to as "toxic knowledge" (Wachbroit, 1996). No wonder only about 12% of people at risk of Huntington's disease choose to have a predictive test (Clarke and Flinter, 1996).

The disability movement needs to reach out to people who receive diagnoses of genetic predispositions, just as it needs to reach out to other groups such as people with HIV/AIDS and older people. Developing alliances with other groups who face stigma and discrimination as a result of their embodiment will be of political advantage in creating a larger constituency demanding change. It will also mean developing a more nuanced and pluralist approach, both to identity and to genetics. The situation for healthy people who are given predictions of future illness will be different from that of people born with impairments or those who suddenly become impaired through trauma. Many of those with genetic predispositions will look to medicine to provide a prophylaxis or cure for their conditions, rather than rejecting medicalization or building a minority group identity.

WHEN IGNORANCE IS BLISS

There are many reasons to consider the new powers of genetic prediction to be a mixed blessing. Knowledge and self-understanding are highly valued in most societies. The truism that knowledge is power is a well-worn cliché, but the corollary that ignorance is bliss may sometimes be a useful counterweight.

This applies to the ways in which medicine deals with information that may be acquired from prenatal testing or predictive testing. Disclosing nonpaternity, for example, can have socially devastating consequences. Perhaps sometimes it would be better not to communicate all the uncertainties, because people may not understand the complexities of the etiology of multifactorial disorders or be able to deal with the calculations of risk involved. Unless we can educate scientists, the media, and the public to communicate genetics in a less deterministic way (see also Kitzinger, this volume), we will face the prospect of people who are diagnosed as having the ApoE4 variant thinking that they are inevitably going to get Alzheimer's or people advised on the basis of genetic testing that they have a higher risk of heart disease fatalistically deciding that they are doomed anyway and therefore can neglect the health promotion advice they have been given.

This may also apply to the ways in which predictive genetics brings new responsibilities for parents and for affected individuals. We need to ensure that we provide support for people faced by difficult dilemmas, not load blame on people who do not act as we anticipate or as we think best. We must accept that people have the right not to know or not to act on their knowledge, particularly given that abortion is a moral issue that divides societies. We must retain a balanced approach to disability, which recognizes that the major problem for many disabled people is the social and physical environment in which they live, not the genes with which they have been born. Our responsibility is to build a barrier-free and welcoming society, not to practice genetic cleansing on the basis of predictions that may often be uncertain. After all, the distinction between disabled people and nondisabled people is breaking down. We all have 100 or more mutations in our genomes, and we all have some degree of susceptibility to polygenic and multifactorial conditions such as cancer, dementia, diabetes, and so forth. The one certain genetic prediction is our mortality, which is written into our very genes. We must try to use the new predictive powers of genetics to enhance human flourishing, rather than to blame the victim and to generate discrimination and inequality.

REFERENCES

Beck-Gernsteim, E., Changing duties of parents: from education to bio-engineering?, *International Social Science Journal*, 42, 451–463, 1990.

Burr, C., *A Separate Creation: The Search for the Biological Origins of Sexual Orientation*, Hyperion Books, New York, 1997.

Clarke, A. and Flinter, F., The genetic testing of children: a clinical perspective, in *The Troubled Helix: Social and Psychological Implications of the New Human Genetics*, Marteau, T. and Richards, M., Eds., Cambridge University Press, Cambridge, 1996, chap. 7.

Cox, S. and McKellin, W., 'There's this thing in our family': predictive testing and the construction of risk for Huntington Disease, in *Sociological Perspectives on the New Genetics*, Conrad, P. and Gabe, J., Eds., Blackwell, Oxford, 1999, chap. 6.

Crow, L., Including all our lives, in *Encounters with Strangers: Feminism and Disability*, Morris, J., Ed., Women's Press, London, 1996, chap. 6.

Driedger, D., *The Last Civil Rights Movement*, Hurst, London, 1989.

Hahn, H., The politics of physical differences: disability and discrimination, *Journal of Social Issues*, 44, 39–47, 1988.

Hall, S., Bobrow, M., and Marteau, T.M., Psychological consequences for parents of false negative results on prenatal screening for Down's syndrome: retrospective interview study, *British Medical Journal*, 320, 407–412, 2000.

Hallowell, N., Doing the right thing: genetic risk and responsibility, in *Sociological Perspectives on the New Genetics*, Conrad, P. and Gabe, J., Eds., Oxford, Blackwell, 1999, chap. 5.

Hood-Williams, J., Goodbye to sex and gender, *Sociological Review* 44, 1–16, 1996.

Jennings, B., Technology and the genetic imaginary, in *Prenatal Testing and Disability Rights*, Asch, A. and Parens, E., Eds., Georgetown University Press, Washington, 2000, chap. 8.

Lippman, A., Led (astray) by cartographic maps: the cartography of the human genome and health care, *Social Science and Medicine*, 35, 1469–1476, 1992.

Marteau, T. and Drake, H., Attributions for disability: the influence of genetic screening, *Social Science and Medicine*, 40, 1127–1132, 1995.

Marteau, T. and Senior, V., Illness representations after the human genome project: the perceived role of genes in causing illness, in *Perceptions of Health and Illness: Current Psychological Research and Implications*, Petrie, K.J. and Weinman, J.A., Eds., Harwood Academic Press, Amsterdam, 1997.

Nelkin, D. and Lindee, M.S., *The DNA Mystique: The Gene as Cultural Icon*, New York, Freeman Press, 1995.

Parens, E., *Enhancing Human Traits*, Hastings Center, New York, 1998.

Rapp, R., *Testing the Woman, Testing the Fetus*, Routledge, London, 2000.

Richards, M., Families, kinship and genetics, in *The Troubled Helix: Social and Psychological Implications of the New Human Genetics*, Marteau, T. and Richards, M., Eds., Cambridge University Press, Cambridge, 1996, chap. 12.

Ridley, M., *Genome: The Autobiography of a Species in Twenty-Three Chapters,* Fourth Estate, London, 2000.

Ridley, M., *Nature via Nurture: Genes, Experience and What Makes Us Human,* Perennial, London, 2004.

Rogers, L., Having disabled babies will be 'sin', says scientist, *Sunday Times,* July 4, 1999.

Royal Association for Disability and Rehabilitation (RADAR), *Genes Are Us? Genetics and Disability*, RADAR, London, 1989.

Shakespeare, T.W., Losing the plot? Discourses on genetics and disability, *Sociology of Health and Illness,* 21, 669–688, 1999.

Shakespeare, T., The dilemma of predictable disablement: a challenge for families and society, in *Conflicts of Interest: Ethics and Predictive Medicine*, Andres, F. and Hay, L., Eds., Institute for Applied Ethics and Medical Ethics, University of Basel, Basel, 2003.

Shakespeare, T. and Watson, N., The social model of disability: an outdated ideology?, in *Exploring Theories and Expanding Methodologies: Where Are We and Where Do We Need to Go? Research in Social Science and Disability,* Vol. 2, Barnarrt, S. and Altman, B.M. Eds., JAI, Amsterdam, 2001.

Spriggs, M., Lesbian couple create a child who is deaf like them, *Journal of Medical Ethics*, 28, 283, 2002.

Tremain, S., On the subject of impairment, in *Disability/Postmodernity: Embodying Disability Theory*, Corker, M. and Shakespeare, T., Eds., Continuum, London, 2002, chap. 3.

Wachbroit, R., Disowning knowledge: issues in genetic testing, *Report from the Institute for Philosophy and Public Policy*, 16(3/4), 1996, http://www.puaf.umd.edu/IPPP/rw.htm accessed January 25, 2006.

Williams, S.J., Is anybody there? Critical realism, chronic illness and the disability debate, *Sociology of Health and Illness*, 21, 797–819, 1999.

8

RACE IN MEDICINE: FROM PROBABILITY TO CATEGORICAL PRACTICE

Richard Ashcroft

CONTENTS

INTRODUCTION

Modern genetics has an uneasy relationship with concepts of race. A singular example of this is the following exchange, which took place at a prestigious Novartis Foundation symposium on the implications of the human genome project for drug development:

> Bradley: I have a question about the number of SNPs [single nucleotide polymorphisms]. Why do we need more SNPs than the number of genes? *Obviously you need different SNPs to define ethnic groups*, but why can't you just have one SNP per 3´ untranslated region?

> Venter: It depends on what your goal in life is. I hope your goal in life isn't just to find genetic variation to identify ethnic groups. That would get us all in trouble pretty rapidly. (Venter, 2000: 16-17, emphasis added)*

Researchers into the biology of responses to drugs have for many years known that variations in drug metabolism can often be explained by genetic variations within and between populations. A branch of pharmacology known as pharmacogenetics has grown up to study genetic variation in drug response, which uses the search for SNPs, among other techniques discussed by Craig Venter and his colleagues in the Novartis Foundation symposium quoted above (Weber, 1997; Rothstein, 2003). A widely shared view in this field is that the distribution of particular variants of genes implicated in these variant metabolic pathways can be mapped reasonably usefully onto ethnic groups, which in turn can be mapped onto geographical regions. The idea is that the frequency of a genetic trait relevant to drug metabolism (as with most genetic variation) will vary between populations and that a particularly important kind of population is the sort that has historically been quite stable.** Some ethnic groups have been stable in this way for one reason or another. A well known example of this kind of stability and its impact on the relative prevalence of genetic variants is the case of certain single-gene disorders known to be highly prevalent in some ethnic groups and rare in others; classic examples are beta-thalassaemia in the Mediterranean population and Tay-Sachs disease in Ashkenazi Jews.*** This is held to make a certain style of essentially racialized medicine respectable in the interests of avoiding inappropriate treatment or serious adverse reactions to drugs (Schwartz, 2001; Robertson, 2002).****

---

* Craig Venter was chair of this Novartis Foundation symposium on pharmacogenetics and was the leader of the private sector human genome sequencing initiative. I have not been able to identify who Bradley is. The symposium proceedings consist of invited papers and verbatim transcripts of discussions, with speakers identified in the original text.

** Stable in the sense that they have not been subject to substantive genetic changes through migration, admixture, genetic drift, or selection.

*** Of course, one could argue that neither the "Mediterranean population" nor the "Ashkenazi Jews" are well-defined ethnic or racial groups. In a sociological sense, the construction of these groups involves a complex blend of social, cultural, scientific, and medical practices, as we shall see.

**** By racialized medicine, I mean a medicine that uses (biological) racial categories in its epidemiology, clinical practice, and therapeutics and that interprets patients' presenting illnesses and behavior on the basis of such categories, through imputation of membership of a racial group on the basis of ethnic origin, cultural background, or skin color and physiognomy or on the basis of patients' self-reported group membership.

This approach to genetic diversity has implications for drug development, clinical trial design and recruitment, and health policy. A particularly important trend in recent years has been the shift from pharmacogenetics (which attempted to identify candidate genes to explain a phenotypic variation in drug response by identifying variations in drug metabolism) to pharmacogenomics (which involves screening the genome for variants in genes known to be relevant to drug metabolism and spotting whether individuals who have a specific variant or set of variants respond differently to the class of drug under test [Mohrenweiser, 2003]). The long-term consequence of this, I shall argue, is a move away from racial profiling to individual-level genomic profiling; the short-term consequence will be that we arrive at this relatively optimistic scenario only by passing through a troubling stage where there is an increased use of racial profiling.

This chapter will discuss the ethical issues involved in this approach to race, genes, and drugs through a critical examination of the ways probabilistic statements about the distribution of genetic variation are used and transformed into categorical statements about genetically distinct groups. We will also discuss the conceptual relationship between the pharmacogenomic approach, often held to offer the promise of individualized medicine, and projects in genetic epidemiology (for instance the HapMap Project). The chapter will conclude by reflecting on what realism about human biological variation implies for medical ethics and whether current constructions of human biological variation are truly realist.

THE LEGACY OF RACIAL THINKING

To understand how what might be quite careful and sophisticated epidemiology trickles down and is received by practitioners, managers, and policy makers, consider the story of Mr. Alton Manning and Mr. Richard Tilt.

Alton Manning was a British black man who died in police custody, having been arrested and subjected to a drugs search. At the time of the arrest he was in good health, and it transpired that Mr. Manning was neither in possession of drugs nor was found to have ingested drugs or alcohol. The *Observer* journalist Nick Cohen reports, in an article titled "Racial Biology", that

> The pathologist found his windpipe had been crushed "due to pressure on the neck." Bruises on his neck, blood spots in the eyes and on the neck and face, dried blood in his ears and mouth, pointed to the same conclusion: a prison manager had grabbed Manning and locked him in a fatal neck hold. (Cohen, 1999:97)

It was a not unheard of case of official violence, then. The twist in the tail of this case was provided by the then director general of the Prison Service, Richard Tilt. As Cohen reports, Mr. Tilt

> brushed aside questions about Manning by saying that "blacks were genetically more likely to be suffocated and killed in custody than whites." He [Tilt] did not seem to understand that blacks are grotesquely over-represented in his jails and that Manning was not killed by faulty DNA but by the brutal and unlawful treatment of private warders. (Cohen, 1999:99)*

It would be convenient if we could take a hard line with such nonsense. It would be helpful if we could say,

> Modern genetics has shown that there is no such thing as a race, so treating people differently, or explaining their differential treatment, on the grounds of race-as-biology is always wrong. It violates the famous Aristotelian formal principle of justice: to treat equal cases equally, and to treat unequal cases unequally. There may be grounds to treat people from different ethnic or sociocultural groups differently in the light of different needs or on the grounds of restorative justice (righting past or contemporary wrongs), but race-as-biology has nothing to do with it, except as part of the discourse implicated in those historic and contemporary wrongs we are trying to correct. Race talk only tells you about the speaker, not the people spoken of (except of being spoken *at* or *about* in this way).

Indeed, this is what we do say — yet still race categories live on through practical daily use (Kohn, 1996; Gilroy, 2000).

Human genetics has not been able to shrug off "racial science" so easily (See also Outram and Ellison, this volume). Human geneticists typically say that there is no genetically defined category corresponding to any phenotypic definition of race and, further, that there is far more genetic variation among individuals who are classified as belonging to the same race as there is among individuals who are classified as belonging to different races, however

* It seems that Mr. Tilt was making some kind of inference from the frequency of sickle cell anemia in the black British population, to a greater susceptibility to asphyxia when restrained. It is not clear that this has any scientific basis, and opinions differ about whether this theory was advanced by Mr. Tilt in good faith. In any case, had he believed it, one might expect different guidelines for the restraint of prisoners and arrestees who are at risk of what he described as "positional asphyxia," yet none appears to have existed. At the very least, this is a case of racializing biology when it suits a public authority and not doing so when it does not.

one defines the races in question or, for that matter, however one defines sameness or difference (Rothstein and Epps, 2001). Nevertheless, beginners' textbooks of medical genetics normally have early chapters with discussions of single-gene disorders inherited in Mendelian fashion such as the thalassaemias, sickle cell anemia, Tay Sachs disease, and cystic fibrosis (Korf, 2000; Rose and Lucassen, 1999), and in these chapters it is normal to find almost casual observations that associate these diseases with particular groups of people defined by ethnic, geographic, or straightforwardly racial categories: "Ashkenazi Jews," "Sardinians," "Cypriots," "Caucasians."* At the same time, such textbooks are normally painfully aware of the history of genetics as embroiled in racial (and racist) sciences and discourses, and of the political minefield in which they tread when seeking to speak of population variation in the prevalence of particular genetic disorders or traits. An indication of this embarrassment is the tangle authors get into when trying to find a language for designating the populations they are discussing — usage of defunct categories from nineteenth century race theory (such as "Caucasian") or concatenations of words (for instance, "race and ethnicity," "racial and ethnic groups"), listed so as not to "get it wrong" and thus offend somebody, whatever "it" is and whoever "they" are (see Pfeffer, 1998 for a commentary on this issue).

RACE AS A MARKER OF GENETIC RISK

There is a big difference between using a set of genetic traits to define a racial category and what will count as membership thereof in society. Likewise, there is a big difference in using a race concept that has not been operationalized biologically as a guide to judging how likely it is that someone might be worth testing to see if he or she is a carrier of a particular genetic marker and is thus suffering from, at risk of, or at risk of passing on to their children, a particular disease. It is quite different to want to define Ashkenazi Jews as a biological category distinct from the rest of us and to want to work with the Ashkenazi Jewish community to carry out Tay Sachs disease carrier screening in their community because members of that community are 100 times more likely to be carriers of the relevant genetic trait than people who are not Ashkenazi Jews (1/3600 vs. 1/360000 [Rose and Lucassen, 1999:35]). Most Ashkenazi Jews see it that way as well — witness the various screening programs and community awareness programs organized within Ashkenazi Jewish communities (Wilfond and Thomson, 2000:64-65). In this case, Ashkenazi Jews and the clinical genetics and public health communities were able to agree that this difference in

* For example, see the excellent and ethically reflective textbooks I use when I am trying to understand genetics from a medical point of view: Korf, 2000:340-3411 and Rose and Lucassen, 1999.

intention and in focus was real and that what was intended was important and beneficial for the target community, and hence a consensus was achieved that, without prejudice to the question of whether it involved a racial element, the screening was not a racist program.*

Philosophers might rest content with what, I hope, is a neatly drawn distinction between the first intention (to define and characterize a racial category through seeking a set of genetic characteristics typically possessed by members of this group and not possessed by nonmembers) and the second (to reduce the incidence of a disease by targeting an ethnic group whose members, as defined by their culture and historical geographical provenance, happen to have a 100:1 relative risk of carrying a particular genetic trait; Boxill, 2001). These intentions are different, from a logical point of view, but how solid is the distinction between them? Is it, as philosophers might hope, a hard-and-fast distinction between conceptual categories or is it something altogether softer, vaguer, and less reliable?

In statistical terms, a risk is a quantifiable probability, based conceptually on the idea of counting events in a definite universe of possibilities: a population.** In practice, this calculation is done statistically by taking samples of the Ashkenazi Jewish population and samples of the total population, and although these figures are not exact, the theoretical basis for the relative risk calculation is as I have given it. Thus, the theoretical basis for the calculation depends on us having (a) a concept of the total number of people in the population (fairly plausible); (b) a concept of the possession of (or failing to possess) the Tay Sachs trait, enabling us to count individuals with the trait (fairly plausible, although it turns out

---

* There is a significant difference, of course, between the politics of engaging with a small, relatively well-known, and politically aware group, such as Ashkenazi Jews, and dealing with continental populations that are diverse and heterogeneous (ironically, considering that from a traditional racial point of view they might be considered more of a whole).

** To state that someone who is an Ashkenazi Jew has a relative risk of 100:1 of being a carrier of Tay-Sachs disease is to say the following: if we take the universe of people alive at some particular time, then we can count them (the number of people is finite). Say there are N of them. Of these, R carry one copy of the Tay Sachs allele; for simplicity assume that all people with two copies of the Tay Sachs allele die at birth (this is not the case; they usually die in early childhood). Thus $N = R + (N - R)$, where $(N - R)$ is the number of people who carry no copies of the Tay Sachs allele. The baseline risk that a person drawn at random from the population will have a copy of the Tay Sachs allele is R/N. Let A be the number of people who are Ashkenazi Jews. Then let R_A be the number of people who are both Ashkenazi Jews and carriers of the Tay Sachs allele. The probability that a given person who is an Ashkenazi Jew is also a carrier of a copy of the Tay Sachs gene is R_A/A. To say that an Ashkenazi Jew has a relative risk of being a carrier of a copy of the Tay Sachs genetic trait is 100:1 is to say that $R_A/A = 100 \times R/N$.

there are a number of different variants of the gene, and the test is neither perfectly sensitive nor specific; Korf, 2000); and (c) a concept of the membership of the class of Ashkenazi Jews (more difficult).

It is of course possible to analyze and unpack any of these concepts sociologically. My claim is that the ease of deconstructing the concept "Ashkenazi Jew" is much greater than the ease of deconstructing the concept of a population or the concept of the Tay Sachs trait, since the concept "Ashkenazi Jew" displays its social constructedness on the surface, so to speak. Acquisition or transmission of the property of being an Ashkenazi Jew is not simply a (biological, heritable) family affair. It depends on the kinship rules of the particular strain(s) of Judaism to which Ashkenazi applies and on the history of the various lineages in the family and where they resided at particular times.* It would appear that to construct a medically useful relative risk, we have to depend on a very complicated and potentially unstable socio-natural construct for our numbers to remain stable and usable. For this to work, one of two things has to happen: (a) either the social relations sustaining the category "Ashkenazi Jew" have to be very stable and robust, and moreover tied to the biological relationships of reproduction and inheritance that we assumed when doing our statistics (which are in fact much more complicated, because they depend on real statistics, not abstract counting in probability space) or (b) somehow the dependence on this type of social relations has to disappear to be replaced by something more robust — in short, we biologize them. If, for example, we can come up with a nice set of biological or medical markers (be they SNPs or family trees) that pick out one-for-one (or at any rate with an acceptable margin of error) who is and who is not an Ashkenazi Jew, then the category will start behaving itself again. Of course, this process is no less social, but the social relations sustaining it are different and often of a less visible and self-disclosing kind.

The balance of this chapter will examine a field in which this kind of rebiologization of a racial (or should that be ethnic?) category (or set of categories) is highly plausible; it will conclude by discussing what, if anything, should be done about it.

PHARMACOGENETICS AND RACE

As most people know, some people suffer very bad allergic reactions to penicillin. Some people get bleeding in their stomachs when they take aspirin; most people do not. Antibiotics, which work well for some people, do not work at all for others. In the 1950s, pharmacologists and doctors

* Compare Marlilyn Strathern's work on the idea that kinship is "purely" biological or "purely" social (Strathern, 1992).

began to recognize that the response people have to drugs is in many cases variable among individuals and that some of this variation depends on genetic variation (Weber, 1997). Specifically, drug metabolism is often dependent on processes in the liver controlled by the secretion of enzymes. Some people metabolize a drug much faster than others (so called "fast metabolizers"). This means they may need to take much larger or sustained doses of the drug than normal because the drug is broken down so quickly that it is excreted before it has any therapeutic effect. Others metabolize a drug much slower than others so that, if they take normal doses, over the usual course of treatment they can experience toxic side effects because the drug, or its residues, builds up in the body rather than being excreted at the normal rate. From the 1950s, instances of drugs for which the response varies across the population in this way have been collected, and in many cases this variation has been shown to be genetically determined. For example, fast or slow metabolizers of a particular drug may have a polymorphism of a particular gene responsible for the production of one of the relevant enzymes responsible for metabolizing the drug. This means that much more or much less of the enzyme is produced or a different variant of the enzyme is produced that behaves in a different way. The study of genetic variations in drug response is known as pharmacogenetics.

The typical research process in pharmacogenetics proceeds from epidemiological observations of patterns of unusual responses to a drug (fast or slow metabolism, side effects unusual in type or extent, adverse reactions, and so on), combined with a search for genetic variation between unaffected and affected individuals that would explain the variation. This research goes alongside pharmacological research into the way the candidate gene is involved in the metabolic pathways of the drug in humans.

From the 1990s, this analytical pathway (from side effect to drug to gene and back) has begun to be reversed with the advent of pharmaco*genomics*. Pharmacogenomics seeks to scan the human genome for genes involved in the metabolism of drugs or for genes involved in disease processes or resistance to disease processes that might be useful in the design of drugs, including the genes of pathogens responsible for disease processes or for the growth, development, reproduction, and immune system of the pathogen (Mohrenweiser, 2003). Using this knowledge, the aim is to design drugs that will mimic the action of resistance-conferring genes; inhibit the expression of genes involved in disease processes, or identify genes involved in drug metabolism (to predict which individuals will succumb to side effects or adverse reactions or will not respond to treatment with a given drug). Pharmacogenomics offers a systematic approach to drug discovery and rational drug design and for avoiding drug-induced injury through individualized prescribing. It has been estimated that between 1 and 3% of all hospital admissions in the United States are due to adverse drug reactions. Drug-induced injury is a major health problem, as well as a major insurance and risk headache for doctors, health services, and the pharmaceutical industry (Kaufman and Shapiro, 2003; Meyer, 2003).

Pharmacogenomics is therefore of enormous interest within public health and the pharmaceutical industry: to researchers at a time when new drug pipelines have been reduced to a trickle while costs of research accelerate; to marketers who will have new drugs to sell and new ways to sell them; and to the lawyers who may be able to get a better grip on product liability risks. Almost all clinical trials of new drugs now involve taking blood samples for banking to use in contemporaneous or future pharmacogenomic studies, but pharmacogenomics also locks in a wide range of other interest groups. Health service managers and managed care organizations are interested not only in reducing their legal risk, but also in cutting expenditure on drugs that do not work for particular patients: once the cost of the test for relevant genes drops below the cost of the drug, it becomes *prima facie* rational to use the test to seek to reduce costs. Researchers in genetics are very interested in pharmacogenomics because it promises to be the only branch of postgenomic science to offer early applications and technological spin-offs, at a time when incentives to enter commercial partnerships are high and public and scientific communities are beginning to ask "So what?" in response to the lengthening series of reports about the sequencing of the human genome. Also, of course, many patients and patient groups are very hopeful that pharmacogenomics will offer treatments for hitherto intractable diseases while reducing the side-effects of existing treatments. The catchphrases for the pharmacogenomics business are "made-to-measure treatment" and "rational prescribing" (Roses, 2004).*

---

* Some commentators on pharmacogenomics have seen the ideas of made-to-measure treatment and rational prescribing as problematic. The threat, as they see it, is a sort of hyperindividualization of treatment and disease: *boutique medicine*. The image of boutique medicine troubles many with the implication that health will become a pure commodity to be purchased in exclusive niche retail outlets, while the consumerization of health care will, among other things, reinforce the links between poverty, ill health, inequities in access to health care, and social exclusion. Further, it risks ever-increasing erosion of the bonds of social solidarity among the sick and between the sick and the well. Some policy puzzles will certainly arise as well. For example, orphan drug legislation, designed to stimulate research and development of treatments for rare diseases and diseases that offer poor return on investment because, for example, they are endemic in the developing world but rare or nonexistent in the developed world, may have the perverse consequence that an incentive exists for microclassifying diseases by genetic variation in treatment response. Since the history of nosology is full of examples where the disease has been defined by the treatment effective for it, this is not a purely theoretical possibility. The outcome of the orphan drug policy could therefore be that orphan drug protections are offered to drugs for so-called common diseases, while leaving the initial problem (i.e. rare or poor world diseases) untouched.

Consider the following scenario. Genetic variation in drug response is not random in the population. Just as in disease genetics, different subpopulations (however defined) have differing relative risks of possessing pharmacogenetically important alleles. Historically, and today, this has been cast in terms of variations between racial or ethnic groups. For example, a family of liver enzymes known as Cytochrome P450 is very important in drug metabolism. Each of the Cytochrome P450 genes has several important variants; one of the most well known of this family of enzymes is known as CYP2D6. This enzyme is important for the metabolism of over 100 known drugs, including codeine, and many important drugs for treating psychiatric disorders and for treating cardiovascular disorders. It is known from pharmacogenetic research that about 6% of White, 2% of African American, and 1% of Oriental patients are poor metabolizers of drugs broken down by CYP2D6 (i.e., they produce this enzyme in an inactive form); 20% of Ethiopian, 7% of Spanish, and 1.5% of Scandinavian patients are ultra-rapid metabolizers of such drugs (i.e., they break them down so fast that huge doses are required). For example, it has been estimated that someone with CYP2D6 deficiency might require a dosage of 10 to 20 mg per day of the antidepressant nortriptyline, while the normal metabolizer would need a dosage of about 75 to 100 mg per day, and an ultrarapid metabolizer would require a dosage of up to 500 mg per day (Wolf et al., 2000:988; British National Formulary, 2000:189). In this case, a general practitioner faced with a clinically depressed Ethiopian patient has a 1 in 5 chance that his or her patient will only respond to a dosage 5 times normal.

This genetic variation associated with different populations presents a number of interesting problems. First, as pharmacologists and geneticists repeatedly point out, little sense can be made at the molecular level of the idea of race (either in its historical formulation or in contemporary notions of racialized ethnic groups). Second, from a clinical point of view, it will be possible in a generation (or possibly sooner) to test people for pharmacologically important genetic variations more or less at the bedside using DNA microarrays (DNA chips). At this point, in theory, the need to decide whether or not to test for such a variant merely on the basis of perceived phenotype, reported race or ethnicity, or clinical history will evaporate. At that point, in theory, all patients will at least stand a good chance of being prescribed at least some drugs that will work for them, at a safe and effective dossage, on the basis of their genotype, rather than their phenotype or their self-ascribed identity or the doctor's assumptions and prejudices — notwithstanding the vagaries of clinical record-keeping.

THE (TEMPORARY?) RESURGENCE OF RACE IN MEDICINE

Until racial and ethnic categories cease to have any operational use, the likelihood is that such categories will have a resurgence in medicine, notwithstanding the acres of print devoted to arguing that they should not be used. This can be expected for a number of reasons:

First, as we have seen, the medical profession is not sophisticated in its handling of racial and ethnic issues, and the racial and ethnic classifications it uses are variable, heterogeneous, and lack a rational basis. Working clinicians and researchers tend to use the classifications that are useful, memorable, and easy to implement. Fairly coarse-grained racialized (that is, race-like) categories, such as those used in censuses, meet all of these criteria. Such classifications can be elaborated and simplified as needed. Where there is no incentive or interest in such elaboration, complication is eschewed. Given that pharmacogenetics is complicated and evolving, and medicine, like history, is "what you can remember," the passage from "in clinical trials, this drug did not work for people with C450 polymorphism X; 20% of African Americans have this polymorphism" to "this drug does not work for black patients" is short and simple (Witzig, 1996; Gordon, 2002).

Second, clinical trials do not always record the ethnic or racial categories of the enrolled participants. Where they do, they tend to use patient-reported or researcher-recorded coarse-grained classifications, rather than anything more sophisticated, as would be required for research in clinical genetics (such as a detailed family pedigree or SNP genotyping). Pharmacogenetic research into drug response typically only takes place if the incidence of adverse events, side-effects, or poor response has an obvious pattern in the population of patients. If the obvious drug response polymorphisms are ruled out, seeking a candidate gene and then its prevalence in the population is difficult, time-consuming, and expensive work. In short, our knowledge of the relevant pharmacogenetics is very patchy, and to some extent what we do know is driven by the tools we bring to the work, such as the categories used in statistical analysis (like the gender, racial, and ethnic categories we have used before).

Third, most pharmacogenomic research is directed at new drug discovery, rather than at further investigations into existing drugs — there being little incentive to invest in such research as patent lives run down. Hence, the knowledge that is used for existing drugs is the knowledge we have, not the knowledge we might ideally want. Moreover, the incentive to seek genetic information about a particular patient in order to make a prescribing decision is relatively low unless the drug is expensive or its toxic side effects are sufficiently dramatic to pose a serious risk. We have, then, a lot of pharmacogenetic knowledge that depends on the coarse-grained racial and ethnic categories used in past (and contemporary) clinical research. A lot of this knowledge is not used or is used at one remove, by which I mean that doctors may save themselves time and trouble by not prescribing a drug that might adversely affect a patient from a group at higher relative risk of adverse events rather than testing to see if this patient actually has the relevant polymorphism (e.g., Satel, 2002).

Sometimes this approach is written into national drug licensing policies. An example is Japan, where, until recently, regulators required trials to be conducted in Japan because of supposed Japanese racial differences in drug metabolism; Japanese recommended dosages tend to be lower than European or American recommended dosages.* This should change as the International Commission of Harmonisation's good clinical practice regulations for interoperability of licensing trials in the United States, Europe, and Japan come into effect, yet there are still many Japanese exceptions permitted by regulations on bridging studies that regulators can require (see Naito, 2000; Nagata et al., 2000).

Fourth, the attention of insurers, health maintenance organizations, and private health care providers in the United States and Europe — and possibly the position of state health services as well (although this is less clear) — will be focused on avoiding unnecessary or ineffective prescribing, iatrogenic injury, and "expensive" patients. While few if any health insurers or care providers are willing or keen to discriminate between patients on racial or ethnic grounds, they are willing and able to discriminate between patients on economic grounds (i.e., if they will be financial risks). If it becomes rational to discriminate, using pharmacogenetic information on economic grounds, this may well turn out to be discrimination on racial or ethnic grounds (because of the association of racial and ethnic categories with economic risk). More to the point, the burden of poor coverage may fall disproportionately on members of particular social groups defined by racial and ethnic categories as used by the medical and administrative staff (and on occasion by patients themselves). This would not be the least bit inconsistent with what is already known about racial and ethnic discrimination in other areas of medicine, such as transplant medicine (Gordon, 2002; Furth et al., 2000), where racial and ethnic variations in Human Leukocyte Antigen (HLA) type form an interesting analogue to racial and ethnic variations in pharmacogenetic polymorphisms, following very similar dynamics to the ones I am tracing here.

It should also be noted that, in the future, the incentive to develop drugs for populations that can both benefit and pay may be intensified; since the incentive will be to develop drugs for the polymorphisms commonest in the richest countries of the world, people with rarer

* While it is true that Japanese patients are more likely to carry certain pharmacogenetic polymorphisms, the prevalence of such polymorphisms is, of course, far from 100%.

polymorphisms or with polymorphisms rare in the rich countries but common elsewhere will continue to be less lucrative markets for such research and development. It is also possible that the more stringent standards for prescribing (including perhaps additional testing to ensure that the patient is eligible for the drug) will raise the cost of treatment even further, making it even less accessible to poor patients in fee-for-service health care systems.

To summarize, adverse drug reactions and the differential effectiveness of drugs can be mapped to genetic variations, while adverse reactions and differential effectiveness are very important for patients, prescribers, and purchasers. It is true that variations by racial and ethnic group have been recorded and can be very important for patients, prescribers, and purchasers. To some extent this is the same as what we saw in the case of disease genetics and differential prevalences of disease-causing alleles in different racial or ethnic populations. However, all of this information is essentially probabilistic, and while in the future it may be possible to screen individuals for the presence or absence of pharmacologically important polymorphisms, this is currently only economically feasible in rare circumstances. For the most part then, the prescribing or purchasing decision will be made on a more or less formal actuarial basis. Hence, racial and ethnic categories are likely to be used more, rather than less, at least over the next 10 years or so, if only in the name of safety and efficacy, rather than (as in the past) in the name of importance or desert.

ETHICAL CONCLUSIONS

My story has been concerned with the somewhat perverse career and peculiar resilience of racial categories in pharmacology and pharmacogenetics. Indeed, just as the Human Genome Project promised to undermine racial biology at the molecular level once and for all, the scientists most involved in the project still felt it necessary to exclude research into the genetics of ethnic groups — at least that is what we are led to think by Craig Venter's comment cited at the beginning of this chapter. What is more, his evasion sounded like a mixture of ethical stance and practical prudence, rather than scientific denial that any such program could or would make sense. Similarly, a large part of the literature on the application of pharmacogenetics in the clinic and in public policy attacks the notion that racial categories could be of any help or any use at all, and much of this writing resonates with the tenor of Venter's remark (for example, Rothstein and Epps, 2001; Schwartz, 2001; Witzig, 1996; Clarke et al., 2001; Epps, 2000; Lewis, 2002; Issa, 2000). Indeed, much of this writing then steps back and posits that although a biologized notion of race is completely unacceptable, for practical clinical purposes an instrumental epidemiological notion of race is both useful and important, at least pending the widespread use of DNA microarrays for routine pharmacogenomic screening.

Many writers taking this line argue that there is an ethical imperative to use such information to benefit patients, particularly, they point out, as minority ethnic health has been given such low priority in the past. Just as therapies developed for women, children, and the elderly are often devised taking male, adult, and middle-aged biology as the norm, non-Whites have typically been treated with medicines designed as if the biological norm was White. Writers could thus argue that a focus on the pharmacogenetic variations between different population groups could lead to improved (rather than unjust and discriminatory) health care for members of historically disadvantaged groups. For example, an important paper by John Robertson summarizes in great detail an argument that using racial categories in this way would be entirely consistent with U.S. constitutional law and would therefore be formally just (Robertson, 2002).

This is an important and powerful argument, which should be taken very seriously. Nonetheless, I believe that the argument I have sketched in this chapter about the series of social and pragmatic factors that lead to increasingly racialized medical practice should make us very cautious about policy design in this area. Not only are there the various good intentions that lead from probability differences to categorical (as in definitive) practice, but there are also the cultural discourses about race as a rhetorical resource for legitimating institutional violence, as in the case of Alton Manning. A good example of the terrain we are operating in comes from a story in the *Financial Times* of March 9, 2001:

> The US Food and Drug Administration has issued an "approvability" letter for a heart failure drug designed specifically for black people, the agency's first formal recognition that different ethnic groups require different medications. The letter paves the way for the clearance of BiDil, a heart failure product made by the small, privately-held US biotechnology company NitroMed. The drug would be marketed only to African-Americans. (FDA Backs Ethnically Targeted Drug, Boston — By Victoria Griffith)

Researchers have long known that certain ethnic groups are more likely to contract some diseases than others, yet it is only comparatively recently that scientists have suspected that population subsets need different medications for the same disease. "Genetic variations between humans may account for their very different reactions to all sorts of drugs," said Mike Hunkapillar, chief executive of Applied Biosystems, which makes the machines used to sequence the human genome.

> Illnesses that seem identical in terms of symptoms — breast cancer or stroke, for instance — may actually be a group of diseases with distinct genetic pathways. This would help explain blacks' far higher mortality rates for a host of conditions, including diabetes, cancer and stroke. Until now these gaps have been attributed largely to racism in the healthcare sector and poverty among African-Americans. At least part of the problem, however, may be that most medications on the market primarily benefit Caucasians. Blacks are twice as likely as whites to suffer from heart failure, and are far less responsive to the widely prescribed hypertension drugs, known as ACE inhibitors. One reason may be blacks' deficiency of nitric oxide, a naturally-produced molecule that dilates blood vessels. BiDil raises the levels of this compound in the body (Hunkapillar cited by Griffith, 2001).

Pharmacogenomics, or individualized medicine, is a key goal of the genomics revolution. Tailoring medications to specific ethnic groups is an important step. "Skin colour may be a surrogate for genetic differences that affect the way people respond to disease and treatment [and] . . . BiDil shows not all drugs are a one-size-fits-all proposition," said Michael Loberg, chief executive of NitroMed, the company behind BiDil. Indeed, the results seem persuasive, as the Financial Times report describes: "Human trials of BiDil showed the drug reduced mortality in 66 per cent of African Americans, but proved of very little benefit to whites." (Griffith, 2001, emphasis added).

Similarly, in an article defending the use of racial categories in diagnosis and treatment, Sally Satel wrote,

> A small percentage of black people with high blood pressure, however, do not have low nitric oxide activity. And the fact that BiDil's intended use relies on a crude predictor of drug response — "a poor man's clue" is how one scientist described race — is something its developers at the University of Minnesota School of Medicine readily acknowledge. *Nevertheless in the sometimes cloudy world of medicine, a poor man's clue is all you've got.* Perhaps that's why members of the Congressional Black Caucus voiced support for the clinical trial. So did the Association of Black Cardiologists, which is helping recruit patients for the trial. (Satel, 2002:15, emphasis added)*

* There was an important debate on this trial and BiDil in the *New England Journal of Medicine* (see Cooper et al., 2003; Burchard et al., 2003; Kaufman et al., 2003), and the debate has continued (see Sankar et al., 2004; and the recently published trial results: Taylor et al., 2004). Details of the trial sponsored by the Association of Black Cardiologists are given in Taylor et al. (2002).

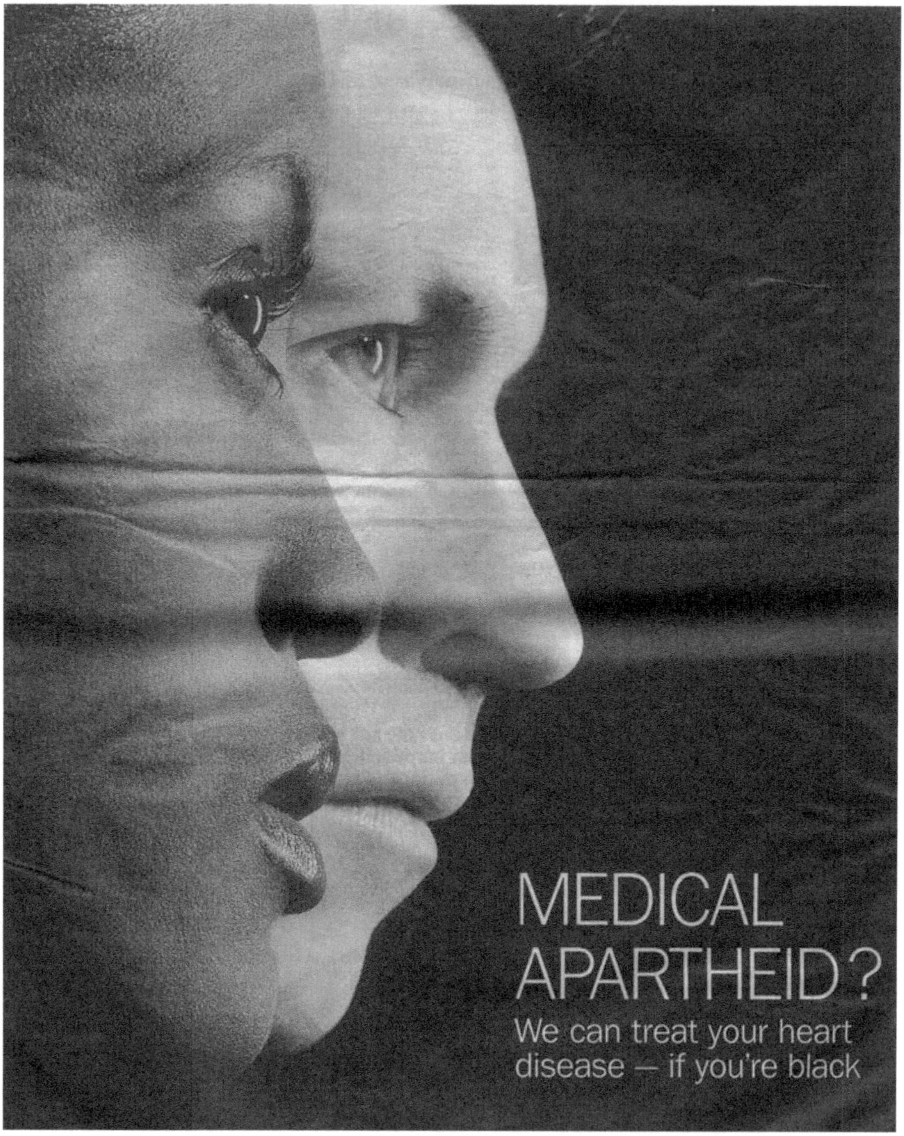

Figure 8.1 How the media greeted reports that a new drug for heart failure ('BiDil') had been successfully tested but only for use on 'Black' patients (*Times*, October 29 2004).

While the "poor man" Satel is referring to is the clinician-as-scientist, one could no doubt come up with other candidates, given the socioeconomic structure of racialized Western societies. Clearly, we should do what we can to remedy inequalities in health. Some of these are biological. Others are due to imbalances or injustices in research and health care budgeting, whether in the past or today. My point in drawing attention to these extended quotations is not to belittle the good intentions of their authors or of the scientists whose work is reported. I want rather to read these reports against the grain to show how they repeat and reinforce the tendencies they superficially stand against.

The ethical implications of the science and sociology of race-talk and race-work in pharmacogenomic medicine are "sometimes cloudy." There is no quick and easy way to better health care, better medicines, and a fairer society. The very least we can do is to challenge the social constructions that underpin structural inequities in our society, insist on justification and explicitness, be alert to social practices disguised as neutral measurement, and resist the shift from probability thinking concerning genes to categorical practice concerning races. This is a labor of engagement, not of producing guidelines.

REFERENCES

Boxill, B., Ed., *Race and Racism*, Oxford University Press, Oxford, 2001.

British National Formulary, *British National Formulary*, 40th ed., British Medical Association and Royal Pharmaceutical Society of Great Britain, London, 2000.

Burchard, E.G., Ziv, E., Coyle, N., et al., The importance of race and ethnic background in biomedical research and clinical practice, *New England Journal of Medicine*, 348, 1170–1175, 2003.

Clarke, A., English, V., Harris, H., et al., Ethical considerations (ethics sub-group, pharmacogenetics working party), International Journal of Pharmaceutical Medicine, 15, 89–94, 2001.

Cohen, N., *Cruel Britannia: Reports on the Sinister and the Preposterous*, Verso, London, 1999.

Cooper, R.S., Kaufman, J.S., and Ward, R., Race and genomics, *New England Journal of Medicine*, 348, 1166–1170, 2003.

Epps, P.G., 2000, White pill, yellow pill, red pill, brown pill: pharmacogenomics and the changing face of medicine, Proceedings from a Conference on The Challenges and Impact of Human Genome Research for Minority Communities, July 7–8, 2000, Philadelphia, PA, http://www.ornl.gov/hgmis/publicat/zetaphibeta/epps.html (accessed January 2006).

Foster, M.W., Sharp, R.R., and Mulvihill, J.J., Pharmacogenetics, race and ethnicity: social identities and individualized medical care, *Therapeutic Drug Monitoring*, 23, 232–238, 2001.

Furth, S.L., Garg, P.P., Neu, A.M., et al., Racial differences in access to the kidney transplantation waiting list for children and adolescents with end-stage renal disease, *Pediatrics*, 106, 756–761, 2000.

Gilroy, P., *Against Race: Imaging Political Culture Beyond the Color Line*, Harvard University Press, Cambridge, MA, 2000.

Gordon, E.J., What 'race' cannot tell us about access to kidney transplantation, *Cambridge Quarterly of Healthcare Ethics*, 11, 134–141, 2002.

Griffith, V., FDA backs ethnically targeted drug, Financial Times, March 9, 2001.

Issa, A.M., Ethical considerations in clinical pharmacogenomics research, *Trends in Pharmacological Science*, 21, 247–249, 2000.

Kaufman, D.W. and Shapiro, S., Epidemiological assessment of drug induced disease, *Lancet*, 356, 1139–1343, 2003.

Khoury, M.J., Burke, W., and Thomson, E., Eds., *Genetics and Public Health in the 21st Century: Using Genetic Information to Improve Health and Prevent Disease*, Oxford University Press, Oxford, 2000.

Kohn, M., *The Race Gallery: The Return of Racial Science*, Vintage Books, London, 1996.

Korf, B.R., *Human Genetics: A Problem Based Approach*, 2nd ed., Blackwell Scientific, Oxford, 2000.

Lewis, R., Race and the clinic: good science?, *Scientist*, 16, 16, 2002.

Meyer, U.A., Pharmacogenetics and adverse drug reactions, *Lancet*, 356, 1667–1671, 2003.

Mohrenweiser, H.W., Pharmacogenomics: pharmacology and toxicology in the genomics era, in *Pharmacogenomics: Social, Ethical and Clinical Dimensions*, Rothstein, M., Ed., John Wiley and Sons, Hoboken, NJ, 2003, pp. 29–49.

Nagata, R., Fukase, J., and Rafidazeh-Kabe, J.-D., East-West development: understanding the usability and acceptance of foreign data in Japan, *International Journal of Clinical Pharmacology and Therapeutics*, 38, 87–92, 2000.

Naito, C., Necessity and requirements of bridging studies and their present status in Japan, *International Journal of Clinical Pharmacology and Therapeutics*, 38, 80–86, 2000.

Pfeffer, N., Theories of race, ethnicity and culture, *British Medical Journal*, 317, 1381–1384, 1998.

Robertson, J., *The Constitutionality of Racial Classifications in Pharmacogenetic Medicine*, University of Texas, School of Law, Faculty Colloquia, Spring, 2002. http://www.utexas.edu/law/news/colloquium/papers/Robertsonpaper.doc (accessed January 2006)

Rose, P.W. and Lucassen, A., *Practical Genetics for Primary Care*, Oxford University Press, Oxford, 1999.

Roses, A.D., Pharmacogenetics and drug development: the path to safer and more effective drugs, *Nature Reviews Genetics*, 5, 645–656, 2004.

Rothstein, M.A., Ed., *Pharmacogenomics: Social, Ethical and Clinical Dimensions*, Wiley-Liss, Hoboken, NJ, 2003.

Rothstein, M.A. and Epps, P.G., Pharmacogenomics and the (ir)relevance of race, *Pharmacogenomics Journal*, 1, 104–108, 2001.

Sankar, P., Cho, M.K., Condit, C.M., et al., Genetic research and health disparities, *Journal of the American Medical Association*, 291, 2985–2989, 2004.

Satel, S., A question of colour, *Guardian*, May 9, 2002, G2 pp. 14–15.

Schwartz, R.S., Racial profiling in medical research, *New England Journal of Medicine*, 344, 1392–1393, 2001.

Strathern, M., *After Nature: English Kinship in the Late Twentieth Century*, Cambridge University Press, Cambridge, 1992.

Taylor, A.L., Cohn, J.N., Worcel, M., et al., African-American Heart Failure Trial: background, rationale and significance, *Journal of the National Medical Association*, 94, 762–769, 2002.

Taylor, A.L., Ziesche, S., Yancy, C., et al., Combination of isosorbide dinitrate and hydralazine in blacks with heart failure, *New England Journal of Medicine*, 351, 2049–2057, 2004.

Venter, J.C., Ed., *From Genome to Therapy: Integrating New Technologies With Drug Development*, Novartis Foundation, John Wiley, Chichester, UK, 2000.

Weber, W.W., *Pharmacogenetics*, Oxford University Press, New York, 1997.
Wilfond, B.S. and Thomson, E.J., Models of public health genetic policy development, in *Genetics and Public Health in the 21st Century*, Khoury, M.J., Burke, W., and Thomson, E.J., Eds., Oxford University Press, New York, 2000, pp. 61–81.
Williams, S.M. and Templeton, A.R., Race and genomics *New England Journal of Medicine*, 348, 2581–2582, 2003.
Witzig, R., The medicalization of race: scientific legitimation of a flawed social construct. *Annals of Internal Medicine*, 125, 675–679, 1996.
Wolf, C.R., Smith, G., and Smith, R.L., Pharmacogenetics, *British Medical Journal*, 320, 987–990, 2000.

PART 3

SCIENTIFIC PRACTICE AND THE PURSUIT OF DIFFERENCE

9

THE TRUTH WILL OUT: SCIENTIFIC PRAGMATISM AND THE GENETICIZATION OF RACE AND ETHNICITY

Simon M. Outram and George T. H. Ellison

CONTENTS

INTRODUCTION

This chapter sets out to understand why racial (and quasi-racial ethnic) categories continue to be used in contemporary genetic research despite sustained criticism from natural and social scientists. It draws on interviews with geneticists working on the editorial boards of 19 high impact genetics journals which routinely publish research using racial or ethnic categories. These interviews suggest that geneticists are generally aware of the questionable reliability, validity, and sensitivity of these categories as markers of genetic variation. However, most geneticists seem to assume that the (modest) genetic differences observed among different populations make racial and ethnic categories useful (if not important) in their research. They also assume that the legitimate use of such categories in genetic research can (and should) be separated from their use as popular components of social identity or as determinants of discriminatory practice within society at large. This chapter argues that this is an essentially untenable position, because geneticists rely on the social salience of race and ethnicity to operationalize these categories as useful markers of geocultural ancestry, while their sociocultural correlates help geneticists understand the differential distribution and penetrance of genetic traits. Indeed, geneticists have effectively subsumed the 'social constructionist' critique of race and ethnicity by using the social nature of these concepts to improve their utility as tools for genetic research. This might explain why geneticists, and others, continue to use racial and ethnic categories *as if* these were reliable, valid and sensitive markers of substantial genetic variation despite evidence to the contrary.

The Social and Scientific Meaning(s) of *Race*

The difficulty with using the term *race* is that it has a range of parallel, formal and informal, theoretical, and empirical meanings (Cashmore et al., 1996). These differ among the natural and social sciences and among clinicians, policy makers, and the public. Indeed, there is even little consensus *within* any of these constituencies. Nonetheless, most of the meanings attributed to the term originate from the contentious idea that the human species can be usefully subdivided into (what some believe to be) genetically distinct races using a range of socio-cultural, geographical, and phenotypic characteristics (Banton, 1998), just as biological taxonomists have used behavior, locality, and morphology to subdivide some animal species into subspecies. Indeed, to many zoologists *race* means subspecies, and some zoologists happily use the terms interchangeably (e.g., Wolf and Soltis, 1992; Krystufek, 2003).

The *idea* that genetically distinct human races exist, is extraordinarily durable and has became part and parcel of the way society at large conceptualizes different human populations and different social groups

(Root, 2003). Over time, race has become associated with a wide variety of different types of group identity: symbolic and material, socio-cultural and structural, phenotypic and genetic. At the same time, the selective allocation of social (dis)advantage along racial lines has made race a peculiarly pernicious dimension of group identity — one with long-lasting material consequences over and beyond its symbolic value as a marker of *imagined* difference. Thus, what was an entirely speculative biological classificatory schema is now associated with very real phenotypic consequences (not least, racial disparities in health [Nazroo, 2003]). As such, 'race' seems to have become, if only to a limited extent, a self-fulfilling prophecy in a process that is worth exploring in some detail.

Racial Classification as a Self-Fulfilling Prophecy

Racial science, an archaic discipline aligned to biological and physical anthropology, sought to classify humans into heritable races or subspecies long before it was possible to directly measure genetic differences (Witzig, 1996). Initially, racial scientists had to rely on the best available alternatives — detailed measurements of phenotypic characteristics (such as cranial morphology and skin color) — to explore human biological diversity. Even when these measurements were conducted carefully, and without reference to the scientist's prior assumptions about which subspecies or race they (thought they) were examining, the measurements could only ever provide crude approximations of genetic variation. This is because the phenotypes examined were all products of interactions between genetic and environmental factors (Cooper and Freeman, 1999). This research tended to focus on comparisons between different geographical populations or different sociocultural populations within the same geographical location — populations that could *already* be distinguished, and appeared very real, if only on sociogeographical grounds. The phenotypic differences observed between these populations and groups were therefore easily interpreted as evidence of both related and unrelated genetic differences: not simply as evidence of differences in the genetic traits (thought to be) responsible for the skin deep morphological differences observed, but as evidence of differences in genetic traits associated with a far wider range of less visible, phenotypic characteristics.

This interpretation turned out to be largely unfounded because phenotypic markers of racial identity do not reflect wholesale genetic differences, and the distributions of racial phenotypes and unrelated genetic traits are often discordant (Relethford, 2002; Livingstone, 1964). Nonetheless, focusing on geographically and socioculturally distributed biological differences in the search for racial subspecies was in many ways an inspired research design. This is because geography and culture are linked to ancestry, and every person's genetic make-up is derived from his or her ancestors (and is more likely to be similar among people who share related ancestors). So, despite inherent methodological limitations in pregenomic racial science, the notion of 'race-as-subspecies,' in both the scientific and social consciousness, was operationalized and became associated with social notions of group identity based on a conflation of geographical origin, culture, visible phenotype, and genetic ancestry. Thus *race* became a word used to describe the geographically and socioculturally related populations being studied by racial scientists as if they were human subspecies, that is, 'race-as-if-subspecies.' In the process, race became one of a range of socially constructed group identities, or *social identities*, and any social identities associated with markers of geographical, cultural, or phenotypic difference became *racialized* and were thereby thought to be related to extensive genetic variation. Defining or operationalizing race and other (racialized) social identities in this way results in them inevitably exhibiting some genetic differences, particularly in the relative frequency of different genetic traits (see Ellison and Jones, 2002). These genetic differences are modest, and we have known for some time that these differences vary more between individuals within groups than between the groups themselves (Lewontin, 1972). As such, they do not permit the classification of genetically distinct human populations that would qualify as racial subspecies, nor do they do much to explain the biological (i.e., the phenotypic) differences observed between populations.

The Consequences of Racial Ideology in Science and Society

Nonetheless, some racial scientists extended the theory of race-as-subspecies way beyond the bounds of careful morphometric measurement and proposed that racial classifications reflected broader genetically determined tendencies in culture, behavior, and temperament. In the process they moved from the legitimate study of phenotypic variability among geographical and sociocultural groups to the development of spurious physiognomic and hierarchical racial classifications. These justified popular racist ideologies in which some human races were viewed as superior to others. The political consequences of this approach to human biology have been well rehearsed elsewhere, while the quasi-scientific claims involved were forcefully rebutted long ago in the postwar UNESCO declarations (UNESCO, 1950, 1951). Nonetheless, the abuses perpetrated in the name of racial hygiene, alongside the sociopolitical and structural consequences of racism, have left a legacy that has profound implications for the meaning of race in contemporary science and society:

First, the plausibility and enduring popularity of racist physiognomies (see Durrheim and Dixon, this volume) means that traditional racial categories are widely believed to provide a genuine, if rough-and-ready, measure of broad genetic difference (Braun, 2002). Moreover, this belief has become subsumed within folk taxonomies and popular scientific explanations of difference (such as Pinker, 2003), which commonly view genetic factors as responsible for biological differences *and* those aspects of behavior that might arguably be considered biologically determined (Tate and Audette, 2001). As such race retains its salience as a group identity associated with biological difference, and in this sense the term has been extended to populations that display comparatively trivial phenotypic or genetic differences (such as Pima Indians and Ashkenazi Jews) in a way that biological taxonomists would never countenance.

Second, the historical impact of racist ideology has created socially heritable structural inequalities in wealth, income, and power among groups classified as distinct races, and these inequalities continue to be reinforced by the enduring popularity of racist physiognomies. As such, race has become salient as one form of group identity associated with the material (including biological) consequences of social (dis)advantage, and in this sense has been applied to disadvantaged populations (such as the Irish in Britain and the Palestinians in Israel) in a way that few biological taxonomists would recognize. Indeed, in a perverse twist of reverse causality, the allocation of structural disadvantage along quasi-racial lines has created environmentally determined social as well as biological differences. These are routinely misinterpreted as evidence that racial classification reflects, rather than creates, biological differences between groups (see the case of Mr. Alton Manning as described in Ashcroft, this volume).

In both senses, whether as a plausible biological ingredient in lay taxonomies or as the biological product of social (dis)advantage, the term 'race' remains relevant to the way in which individuals and groups experience their social and biological identity. As such race can be used to classify populations into racial groups (even though these are arbitrary groupings and poor approximations of subspecies) and to operationalize biological differences between racial groups (even though these differences are consequences rather than determinants of racial classification). Likewise, the racialization of intrinsic and extrinsic biological variation among other social identities means that these identities can also operate as quasi-biological categories, even when they were originally conceived as predominantly social entities. Perhaps the most important of these is ethnicity, a form of social identity associated with sociocultural factors and often operationalized as a subset of race (Friedman et al., 2000). Like race, ethnicity is a socially constructed group identity that appears biological when it is defined or measured using biogeographical characteristics (such as geographic origin, ancestry, heritage, or phenotype), that is, characteristics more commonly attributed to race (Smaje, 1995). As such, ethnicity operates in a similar fashion to race and can provide a plausible proxy for a similarly modest amount of genetic variation while avoiding an explicit association with the controversial and discredited theories of racial biology.

Why Do Geneticists Use Race and Ethnicity?

Like any discipline using population-based analyses, such as epidemiology or demography, genetics has had to rely on geographical and sociocultural characteristics to identify, disaggregate, select, and compare human populations. Without recourse to phenotypic or genotypic measurements on entire populations — a time-consuming, costly, and impractical exercise — geographical and sociocultural characteristics have been the only "handles" available for operationalizing populations as analytical units. It is, therefore, unsurprising that racial and ethnic categories have been embraced as analytical tools by population scientists (including geneticists). However, such research has tended to *essentialize* the classifications involved (Pfeffer, 1998) whatever the subject of enquiry (be it sociocultural, behavioral, structural, or biological) in a process that has been called 'racialization' (Goodman, 2000). And when the subject of enquiry is genetic, the essentialization of such classifications is compounded by what Lippman (1992) has called 'geneticization' (akin to genetic determinism or genetic reductionism) – the impression that the classifications and their associated characteristics are genetically determined or can be reduced to genetic factors. Thus when populations examined for genetic analyses are classified or labelled using racial (or quasi-racial ethnic) categories, racialization combines with geneticization to essentialize the categories as genetic (Goodman, 2000).

While geographical gradients (i.e., clines) in genetic variation and some sociocultural differences in genetic relatedness (such as those associated with descent or kinship) are widely accepted and ostensibly acceptable facts within science and society, the use of race and ethnicity as markers of these genetic differences remains highly contentious. Despite claims to the contrary from those in charge of the Human Genome Project (see Ashcroft and Goodman this volume; and McCann-Mortimer et al., 2004), one unforeseen impact of the new genetic technologies has been to emphasize the genetic differences between racial and ethnic groups, even though much of the research to date has found these differences to be modest and ephemeral (e.g., Rosenberg et al., 2002). Although such studies do find some genetic differences between groups classified using traditional or contemporary racial (and related ethnic) labels, particularly in the frequency of genetic traits shared unequally among isolated populations, they do not prove that race is genetically determined. Instead they demonstrate how what is actually a highly contested idea (that genetically distinct human races exist) can be taken

Figure 9.1 Despite sustained criticism from natural and social scientists, 'race' remains an important focus for enquiry in genetic research, as evident in this 2004 special issue of the influential academic journal *Nature Genetics*.

up as a marker of *social* difference and thereby create *social* groups with modest genetic correlates and consequences. In other words, the genetic differences observed among different populations do not make these races, but in the way these populations are classified as races, they are made up to be genetically different (Taussig et al., 2003).

Clearly the new genetics has a seminal role to play in the way we now perceive race, ethnicity, and other social identities (Braun, 2002), yet the processes underpinning the tenuous associations between social identity and genetic variation remain fundamentally social. The use of these associations in genetic and genomic research is therefore likely to have a number of *naturalizing* effects within science and society (Ellison and Jones, 2002). In particular, the geneticization of social identity has the potential to naturalize biological and social inequalities that are the consequence of the historical and contemporary allocation of social (dis)advantage such as inequalities in health along racial and ethnic lines (Nazroo, 2003). As such, their use in genetic research has come under sustained attack from those who argue that, as social constructs, race and ethnicity have limited reliability as scientific variables and limited validity as markers of genetic variation. Moreover, there is increasing concern about the social and political sensitivity of using these concepts in genetic research, where they have the capacity to reify discredited notions of innate difference and feed into racist ideology (UNESCO, 1997). The use of racialized categories in genetic research therefore poses a dilemma that is part scientific and part ethical. Yet to a large extent the use of race, ethnicity, and related social categories remains routine practice in much genetics research.

The aim of this chapter is therefore to focus on the particular concerns relating to race and ethnicity as analytical concepts in genetic research and explore why geneticists continue to use racial and ethnic categories in their research, drawing on the experiences and views of genetics journal editors as potential arbiters of good practice and experienced researchers in their own right. The chapter uses qualitative interviews with genetics journal editors to examine the formal and informal practices governing the collection and presentation of genetic information disaggregated by race or ethnicity.

METHODS

Ellison conducted 22 semi-structured interviews with geneticists working on the editorial boards of the 19 most highly cited genetics journals publishing research using racial or ethnic categories (see Ellison and Jones, 2002: Figure 1, p. 266). Given the geographical distribution of these journals' editorial offices (13 in the United States, 4 in the United Kingdom, and 2 elsewhere in Europe), only one of the interviews took place face-to-face, and the remainder were conducted by telephone. Consent was obtained from each interviewee to tape-record the interview, following an assurance that the interview would be analyzed anonymously. These recordings were later transcribed to facilitate analysis and the extraction of verbatim quotes to illustrate the key themes that emerged.

The interviews lasted between 50 and 120 minutes and covered four broad areas using a flexible topic guide containing a range of questions and prompts prepared following informal discussions with both geneticists and anthropologists. These four areas examined were: (a) the nomenclature, classification, and measurement techniques used to operationalize race and ethnicity in genetic research; (b) interviewees' views regarding the reliability and validity of race and ethnicity as genetic and sociocultural variables; (c) the principal explanations offered for racial and ethnic variation in the distribution and penetrance of genetic traits; and (d) possible options for strengthening the use of race and ethnicity in genetic research. The analyses that follow draw on all four of these areas, paying particular attention to the first three.

In the absence of visual cues for sustaining dialogue, the telephone interviews sought to generate free-flowing conversation between interviewee and interviewer. This encouraged the interviewees to explore avenues of personal interest, experience, and expertise rather than restricting them to issues on the topic guide. This approach generated rich and diverse data appropriate for analysis using *grounded theory* (Glaser, 2001; Strauss and Corbin, 1998) — a technique that allows core themes to emerge from the interviews, drawing on explicit statements made by the interviewees rather than any prior expectations of the researchers and interviewers. For similar reasons, *discourse analysis* (Wetherell et al., 2001) was used to explore the implicit reasoning behind different points of view and to examine how each of the interviewees presented their own experiences and expertise when constructing their scientific and social explanations. These analyses involved both authors (S.M.O. and G.T.H.E.) listening to each recorded interview, and repeatedly reading each transcript, to identify and reach consensus on the significant themes and rhetorical devices evident therein.

RESULTS

Do Geneticists Recognize that Racial and Ethnic Categories Are Unreliable, Invalid and (In)Sensitive?

All of the geneticists interviewed found it difficult to offer a clear definition of race and ethnicity and to distinguish between the two: "How one classifies a group is . . . a very good question, and the answer is extremely complex." Most offered ad hoc suggestions for operationalizing race or ethnicity using phenotypic, geographical, and sociocultural criteria. For example, one interviewee described how "if I were to use race . . . I would define it as a word that encompasses a group of people morphologically and, I would hope, genetically more related." Others felt that ethnicity was "synonymous with nationality" or that "most people [researchers] use skin colour," while some suggested that religious affiliation or a combination of criteria was best: "something as simple as skin colour and self-declaration of ethnicity, with some confirmation of parents' and potentially grandparents' [origins]." Clearly there was little consensus on how race and ethnicity should be defined or measured in genetic research. There was even some recognition that the categories and nomenclature used are subjective, context-specific, and varied over time.

The questionable reliability of these categories was therefore implicit in the varied definitions and approaches to measurement and in the view that they could be "used interchangeably." It was also explicit in the comments of a small number of interviewees who described racial and ethnic categories as "simplistic," "imprecise," and "sloppy" and who felt that current practice is quite simply "a mess." When asked to distinguish between race and ethnicity, one of these interviewees described how they "could find essentially the same definition meaning both things, and there were a number of them that could sometimes be applied to one and sometimes to the other — so it was very, very murky." Indeed, the absence of what was described as a "community standard" for measuring race or ethnicity led a few of the interviewees to call on journals to "establish appropriate levels of information" and on editors to "play a major role" in promoting these.

While the varied definitions and approaches to measurement demonstrate that racial and ethnic categories are inherently un*reliable* social constructs, only a few of the interviewees seemed to explicitly recognize this. In contrast, most acknowledged that these categories have limited *validity* as scientific "surrogate[s] for genetic identity." Although some interviewees sought to draw a distinction between race and ethnicity, suggesting that "one [race] is determined in my mind strictly by genes, and the other [ethnicity] is determined by a combination of genes and heritage and environment," most recognized that operationalizing these using phenotypic criteria or self-reported identity undermined the genetic precision of both. Thus, many shared the view that "there isn't always a clear correspondence, certainly at the individual level, between self-identification, labels of ethnicity, group membership, and broader patterns of genetic group.

Likewise, the interviewees acknowledged that phenotypic criteria reflect "a combination of genes and heritage and environment" and that "without that information [on heritage and the environment] the measured phenotype may be wrongly ascribed to a particular genetic determinant where it could easily be environmental or some other determinant." The greater level of consensus among interviewees on this issue probably reflects their common scientific expertise, as one of the interviewees explained: "I am familiar, as a geneticist . . . with the superficiality of the term 'race'." Indeed, many interviewees were able to offer cogent examples of the etiological difficulties created when biological differences between racial and ethnic groups were interpreted as genetic. For example, one asked, "Is it that your dietary habits affect your risk of prostate cancer? Is it that your care and early detection is not as good because you're Black and more likely to live in a poor neighbourhood? Is it genetics? What is it?" The interviewees were, therefore, not convinced that the criteria used to classify racial or ethnic groups provide the most appropriate markers of genetic variation, and many agreed that "which criteria are related to genetics is something that should be addressed."

Finally, all of the interviewees seemed acutely aware of the social and political sensitivities associated with using racial and ethnic categories in genetic research. These sensitivities concern the use of archaic and potentially offensive nomenclature; the impact of genetic findings on public notions of identity; and the tendency for genetic traits to stigmatize those groups in which the traits are most common. Many recognized that traditional racial nomenclature is "problematic" and "more confrontational" than contemporary ethnic labels, although some appeared perplexed by the changing fashion for particular terminology. For example, one interviewee described how "at one point it was absolutely politically incorrect to refer to 'Blacks' — and now it's not." Likewise, others seemed surprised by the public's "peculiar fixations about where their genes come from" and how these have "a profound effect on their feelings of identity." Nonetheless, some recognized that genetic research simply has "a human interest angle" that makes genetic information particularly salient and powerful, not least when associated with contested group identities. Indeed, one of these interviewees seemed astonished that geneticists fail to grasp the social and political significance of their work: "It seems to me just a kind of naïveté that geneticists think that this kind of information won't be of import to individuals." Even those who justified the use of racial and ethnic categories to identify and eliminate genetic diseases recognized that stereotyping is a serious potential consequence. Thus, when describing how prenatal screening for common genetic diseases among Jewish communities had reduced the incidence of these conditions, one interviewee conceded that "the other side has been that now . . . they are associated with these diseases."

Clearly, the geneticists interviewed view racial and ethnic categories as unreliable, invalid, and politically sensitive tools, even though they did not always acknowledge this or find it easy to articulate why. A few were openly critical of using these categories in genetic research and a minority were keen to develop and apply new approaches for classifying populations that have more relevance to genetic research. However, most adopted a laissez faire approach, opting for more detailed reporting of the classificatory methods researchers use in the belief that the contested value of using racial and ethnic categories in genetic research would only be resolved by the results of this research.

How Do Geneticists Justify the Use of Racial and Ethnic Categories in Genetic Research?

It was certainly surprising that so few interviewees were keen to stop using race and ethnicity, not only because most had difficulty describing how to define and measure these, but also because they were all aware that using these terms made them subject to public sensibilities. Despite widespread criticism of the validity of racial and ethnic categories as "surrogate[s] for genetic identity," the principal justification for continuing to use these categories in genetic research was that their utility is borne out by the results. Indeed, more than one interviewee argued that the "very good evidence" of biological differences between racial and ethnic groups meant they had a "social, moral and political obligation" to continue using these categories and to report, rather than "finesse," any differences they found. In these arguments disparities in health played a special role, both as evidence of biological difference and to demonstrate the clinical relevance of their work: "We don't want to stereotype, we don't want to profile, you know. On the other hand we don't want to ignore the fact in epidemiology that there are diseases of gender and ethnic background that are in one group but not another."

Interviewees made particular use of symbolic Mendelian diseases that have well-established associations with specific ethnic groups (such as Tay Sachs and sickle cell trait) to emphasize the contribution of genetic factors to disparities in health. They drew parallels between these disparities and the genetic trends they found elsewhere and tended to overlook or play down the relevance of environmental factors in the more complex diseases involved. In this way even those interviewees who had openly criticized the quality of measurement practice and the genetic validity of racial and ethnic categories constructed a robust argument that "some types of groups" are valid markers for some genetic traits. For example, one felt that although the categories "are self-described and poorly defined by anyone's standard . . . you still see grouping together at the genetic level," while another argued that "the [genetic] differences we are measuring between different ethnicities are greater than the error that we are incurring in not classifying certain people correctly." Thus, racial and ethnic groups were transformed from rough and ready categories into "very useful handles on population differences," and despite widespread uncertainty about how these should be defined and measured, there was substantial support for the view that "the community of population geneticists have already worked through what categories are valid in the research."

This view, that geneticists have the requisite collective experience to establish the validity of the analytical categories they use, touches on a key rhetorical device the interviewees used to justify the continuing use of race and ethnicity in genetic research: separating scientific expertise from lay knowledge and public (mis)understanding. This device set geneticists and genetic practice apart from society and emphasized the difference between the use of racial and ethnic categories as scientific tools in genetic research and their role in popular notions of social identity. Indeed, the interviewees were keen to point out that the public has a very different understanding of the link between genetic variation and identity, one describing an instance where they had been approached by "a woman who claimed to have gypsy ethnicity and wanted us to prove she was a gypsy using genetics . . ." and who became "very angry . . . when we said that this was impossible . . .."

While the popular meanings of race and ethnicity were characterized as very different to the meanings attached to these when used as scientific tools in genetic research, the public understanding of genetic research was also felt to be "confused." Thus, another interviewee described how "it is often difficult to make clear [the] distinction between the scientific interpretation of race and the social interpretation of race and they get clouded quite often — get confused in the public's mind." In this way the interviewees largely accepted that race and ethnicity have a range of different meanings — scientific and societal, biological and social — and to some extent recognized that the different meanings serve different purposes in each of these contexts. For example, when discussing legislation mandating the collection of racial and ethnic data in the United States, one interviewee described the classification used as "very far away from the biology." However, most interviewees felt that much of the sensitivity surrounding the use of racial and ethnic categories in genetic research is compounded by the very different lay meanings attributed to these and by public misunderstanding of genetic findings. Just as interviewees seemed perplexed by the changing fashion for what constitutes "politically correct" terminology, they often seemed exasperated by having to defend their scientific work: "One has to be careful to not become part of a political process but rather focus on communicating research findings."

How Do Geneticists Accommodate the Argument that Race and Ethnicity are Socially Constructed?

It appears that the separation of science from society serves two functions: to strengthen the credibility of racial and ethnic categories as robust and legitimate scientific variables (when used as such in genetic research) and to avoid contentious public debates (including accusations of inaccurate, insensitive, or inflammatory research). As a rhetorical device, however, the separation of genetic research from its wider social context is essentially untenable — not least because some of the interviewees drew on the social salience of race and ethnicity to operationalize these as self-reported scientific variables and cited the potential utility of such categories in clinical and related social contexts to justify their scientific importance. Indeed, all of the interviewees recognized that race and ethnicity have a number of social attributes that need to be accommodated when operationalizing and justifying them as scientific variables, even though few described them as 'social constructs' and none embraced the view that racial and ethnic categories cannot be divorced from the social processes that produce them. Instead the interviewees adopted what appeared to be an entirely pragmatic stance in which the social attributes of race and ethnicity were viewed as both challenges and opportunities. This was evident in discussions about: the choice of nomenclature and terminology; the use of social determinants and correlates of group identity to improve the acuity of genetic research; and the impact of social policy on research practice.

Thus, although many of the interviewees felt that race is a more appropriate genetic category than ethnicity (which many felt was "determined by a combination of genes and heritage and environment"), there was general agreement that "race has become tainted, much more so than ethnicity, so people tend to prefer using ethnicity." Although interviewees were uncomfortable with the inherent subjectivity of self-reported identity and recognized that "self perception of group membership or ethnicity may or may not correspond with what some genetic analysis says," they nonetheless recognized that self-reports were increasingly viewed as the most acceptable approach to classification. In this way the interviewees took what might be described as a politically expedient approach — using whatever categories, labels, and measurement techniques were socially acceptable, while distancing themselves from the limited reliability and validity of this approach by arguing it had "nothing to do with science." As such, the interviewees seemed to be suggesting that social forces are responsible for the categories they are obliged to use *and* for their limited acuity.

Nonetheless, when discussing the social determinants and correlates of group identity, some interviewees saw these as potentially valuable analytical tools: first, because information on "population history" might reflect some of the social and demographic processes that contribute to genetic variation and second, because data on social correlates might help address the impact of environmental confounding on phenotypic variation. For example, one interviewee suggested that "if we can layer [on that] our knowledge about ancestry of the population, how people are related and likely to be related to each other, we can gain a lot more information, a lot more power." Another described how "there's been lots of discussions saying 'how can we build demographic and ancestral information about maybe migration and population structures into our studies to get more genetic information out of it?' I think it's achievable. That would probably be a big improvement to the field." Some specific examples were given to illustrate this approach, such as recent studies in Finland, where it was felt that local researchers "know their history: that originally people moved into the South, then into the North. This knowledge of that history has been essential to the design of genetic studies in that population." Meanwhile, some interviewees readily accepted that they were largely uninterested in the nongenetic aspects of race and ethnicity and wanted "to try to strip out [everything else] and just look at the genetics." To this end one interviewee thought that race and ethnicity could be used to help "see associations through the noise, in terms of stratifying to sensibly take into account of the noise, whatever that kind of noise might be — environmental/background."

Clearly the social aspects of race and ethnicity were not simply viewed as inevitable limitations of unscientific social characteristics, but were also felt to be potentially useful for scientific analyses. In this way the interviewees essentially co-opted the social aspects of race and ethnicity, using them to justify the use of categories with limited scientific reliability or genetic validity and to explain both the genetic and phenotypic differences they found. When all else failed, interviewees from the United States often invoked recent changes in NIH* policy, which requires the projects it funds to collect data on the racial and ethnic affiliation of research participants. These interviewees recognized that such data were intended to ensure that all groups were included in, and benefited from, research, but some saw this as "a mandate" to include racial and ethnic analyses in their research even though these comparisons involved questionable categories and were potentially (in)sensitive. Interviewees from elsewhere used similar arguments regarding the routine collection of racial and ethnic

* The NIH is the National Institutes of Health, which describes itself as "the Federal focal point for medical research in the United States . . . comprising 27 separate Institutes and Centers . . . [and] one of eight health agencies of the Public Health Service which, in turn, is part of the US Department of Health and Human Services." (NIH, 2005).

data for monitoring purposes to explain the continuing availability, and use, of these data for genetic research. However, for the most part, the interviewees preferred to justify the continued use of racial and ethnic categories on ostensibly scientific grounds, and this explanation was usually invoked as simply the ultimate *fait accompli*.

Potential Limitations

Before laying too much store by this analysis of the views and experiences of just 22 genetics journal editors, it is important to bear in mind a number of potential limitations and biases. Like any qualitative study using a purposive sample, its findings cannot claim to be representative of all genetics journal editors, let alone geneticists more generally. Indeed, by focusing exclusively on genetics journals that have published research using racial or ethnic categories, this study aimed to engage with editors conversant with the way these categories are operationalized in genetic research. As such, the views and experiences analyzed here may be better informed than those of editors and geneticists working in contexts where racial and ethnic categories are irrelevant or rarely used. To some extent this was evident in the different views and experiences of editors with different backgrounds and those working on different types of journals. For example, some of the editors were not specialists in human genetics, and some of the journals had published a range of human and nonhuman studies. A few of the editors had training in one of the population sciences (such as anthropology or epidemiology) as well as genetics or worked on journals specializing in human population genetics. In the main, the latter felt they were more knowledgeable about, and were more critical of, the use of racial and ethnic categories than the former. This might explain the lack of any clear consensus on such issues as the definition and genetic utility of race and ethnicity, but it might also mean that the analyses better reflect the range of opinion across the community of geneticists than might have been expected from the rather narrow, purposive sampling frame.

This variation might also provide some reassurance that the interviews avoided a potentially serious source of bias related to the interviewer's critical stance on racial and ethnic classification in biomedical research (Ellison et al., 1996, 1997) and to the way in which questions about "strengthening" the use of race and ethnicity in genetic research positioned the study as critical of current practice. To address this issue the interviewer sought to focus on both the scientific merits and scientific problems of using racial and ethnic categories in genetic research, rather than the social sensitivity to these, and encouraged interviewees to apply their scientific expertise to reflect on how these categories were operationalized and interpreted. While a few interviewees appeared cautious and defensive, focusing their comments on the discredited history and contemporary social sensitivity of racial and ethnic categories as opposed to their potential scientific utility, most appeared to freely discuss their scientific views and experiences. This suggests that most of the interviewees offered views that were not constrained or unduly influenced by the sensitivity of the issues discussed. However, many suggested that the interviews had encouraged them to reflect on beliefs and practices they had paid little attention to before, and it may be that that some of these interviewees expressed views and arguments that they had not thought through. As such, it is likely that many of the interviewees could have offered more considered responses given the opportunity to reflect on the issues discussed, and that the views emerging from the interviews were occasionally somewhat unformed.

Notwithstanding these limitations, some clear themes emerged from the interviews and reflect a number of explicit and implicit justifications for the continued use of race and ethnicity in genetic research.

DISCUSSION

Few of the interviewees seemed to recognize that race and ethnicity are fluid and contingent concepts and are, therefore, inherently unreliable as scientific classifications of fixed categories with discrete boundaries. This seems remarkable given the lack of consensus on how either should be defined or operationalized, which both reflects and contributes to their limited reliability. While there was some consensus on the limited acuity of race and ethnicity as markers for genetic variation, this was largely ascribed to the discordant distribution of genetic traits rather than the variability of the classifications used. Indeed, for many of the interviewees the patterns they observed for some genetic traits across some populations validated the utility of racial and ethnic categories (in some circumstances) and overcame any lack of measurement precision. This apparent anomaly is partly semantic and partly epistemological: discussing the genetic acuity of racial and ethnic categories requires, and demonstrates, that they can be articulated *as if* they exist; assessing the genetic acuity of racial and ethnic categories requires, and demonstrates, that they can be measured *as if* they are reliable. Although race and ethnicity are fluid and contingent because they are socially constructed (Tatum, 1997), discussing or operationalizing them *as if* they were reliable means they can be mistaken for natural entities.

This has a number of consequences for scientific research: any characteristics associated with race or ethnicity become essentialized, that is, they appear to be integral components of the populations involved, and any researchers exploring such associations are vulnerable to accusations of racism or ethnocentrism — particularly when the characteristics involved have the power to further essentialize or stigmatize the populations involved. These problems are, perhaps, most acute for genetic research because some genetic traits can easily be presented as making a fundamental contribution to human characteristics — making geneticization a particularly powerful form of essentialization (Nelkin and Lindee, 1996). Although an emphasis on the clinical utility of research into the genetic determinants of disease may help the discipline rise above its association with the eugenic abuses of the past, identifying the genetic traits responsible for diseases that are more common in specific populations can stigmatize these groups as intrinsically unhealthy.

The interviewees were clearly aware that race and ethnicity are sensitive concepts capable of generating substantial public controversy, but while they were keen to avoid such controversy, few were prepared to forego the use of racial or ethnic categories in genetic research — at least not until there was more conclusive evidence of their questionable utility. Instead they sought to distance their work from the wider social context in which this took place and put the controversy down to the public misunderstanding of science. Yet these proved to be essentially untenable strategies because contemporary norms of acceptable scientific practice require that race and ethnicity are operationalized using socially salient and socially sanctioned classifications — that is using categories and labels that are recognizable, and recognized, as legitimate and acceptable even when they have limited validity as markers of genetic variation. These are not necessarily the categories or labels geneticists would ordinarily choose, not least because their salience as markers of social identity means they are likely to be associated with a host of related social, structural, and environmental variables capable of confounding any relationship between genetic and phenotypic traits. Moreover, recent advances in genetic research have generated intense public and media interest. The importance of explaining these findings (and justifying the costs involved in generating them) means that geneticists have to communicate with the public, and with those who fund their research or translate their findings into products, services, or policy, in terms they understand. Thus, geneticists cannot dissociate the use of racial and ethnic categories as scientific tools from the social context in which these are produced, and while the categories and labels they are obliged to use are increasingly subject to social control and public scrutiny, these can help improve the translation and application of genetic findings to other contexts.

Given their readiness to separate their science from society to deal with the problematization of race and ethnicity, it was somewhat unexpected that the interviewees would also offer such a pragmatic response to the social constraints they faced when operationalizing these concepts as scientific variables. To some extent this simply involved accommodating these social constraints, first by adopting labels and nomenclature that were deemed socially acceptable and second by accepting that, as social categories, these would have variable genetic acuity. Thus, while most of the interviewees preferred the term *race* (with its biological connotations) to *ethnicity* (with its cultural connotations), they generally accepted that both are social concepts with limited genetic acuity and were prepared to use *ethnicity* as a more socially acceptable term than *race*. In the process both terms were essentially redefined: *race* was no longer seen as a natural entity determined by distinct genetic differences but as a social term, not dissimilar to *ethnicity*, with looser genetic affiliations; *ethnicity* was no longer seen as only a fluid version of group identity that captured sociocultural diversity, but also as a marker for ancestry and thereby associated with potentially useful amounts of genetic variation. On the one hand, then, geneticists have simply accepted the criticisms leveled at *race* (as an unreliable, invalid, and insensitive measure of broad genetic difference) while, on the other hand, they have embraced *ethnicity* (as a quasi-racial social concept that provides a more acceptable, if looser, marker of associated genetic variation).

Moreover, the interviewees suggested that geneticists are keen to make a virtue out of necessity by using the sociocultural correlates of ethnic (and racial) categories to improve their value as markers of genetic difference. This involved substantial interest in capturing those social and historical characteristics of populations that reflect their ancestry and genetic heritage. It also led the interviewees to reflect on the potential use of these categories to control for sociocultural confounding in any relationships observed between genetic and phenotypic traits. While none of the interviewees appeared to recognize the value of socially salient nomenclature in translating their findings into social contexts (such as the media, the clinic, and the market), this may simply reflect the predominantly scientific focus the interviews took, particularly since the translation and portability of genetic findings (from bench to bedside) has featured prominently in genetics journals (e.g., *Nature Genetics*, 2004).

The integration of the social constructionist critique of race into the operationalization of ethnicity as a quasi-racial genetic variable is not an example of "ontological gerrymandering" — a phrase coined by Woolgar and Pawluch (1985) to highlight the partiality of social scientific debate. However, the term does seem appropriate for the way in which geneticists have essentially co-opted aspects of the critique that help them justify their continued use of racial and ethnic categories. Certainly, in the way they selectively cited Mendelian traits that vary among racial or ethnic groups as evidence of wider genetic differences, the interviewees demonstrated a readiness to forgo impartiality. This was also evident in their focus on genetic differences between isolated populations, which tends to accentuate the extent of geographical differences in genetic variation (Outram and Ellison, 2005). While they drew on what Gilbert and Mulkay (1984) describe as "empiricist repertoires" to separate the objective scientific use of racial and ethnic categories as rational from their "contingent" – nebulous and subjective – social meanings, their principal justification for continuing to use these categories as genetic markers stemmed from a fundamental confidence in science. This involved what Gilbert and Mulkay (1984) have called the "truth will out device," or TWOD, through which "gradually, over time and given enough sound empiricist research, the truth of the natural world would reveal itself" (McCann-Mortimer et al., 2004:415). To this end, the interviewees cited instances where genetic differences were evident between different racial or ethnic groups as evidence that such differences *did* exist and *should be* explored. Moreover, this evidence encouraged the interviewees to argue that it was premature to reject the use of racial or ethnic categories in genetic research until the extent, and value, of these categories had been fully explored.

Taken as a whole, these strands of reasoning provide a relatively coherent and convincing argument, not least to geneticists, who are primarily interested in the scientific utility of race and ethnicity rather than their social or political complications. Little incentive therefore exists for geneticists to stop using these categories in their research or to constrain the use of these in scientific contexts (Outram and Ellison, 2006), provided they address the potential sensitivity of the nomenclature they use (Sankar, 2003) and are prepared to accept that the categories only provide crude markers of genetic difference. Even if geneticists adopt this more cautious stance, it seems likely that the continued use of racial and ethnic categories in genetic research will inevitably strengthen the impression that these are "fixed" genetic categories (Goodman, 2000; and see Ashcroft, this volume). Indeed, as Gannett (2001) has argued, "group DNA differences will always be culturally meaningful and socially relevant — even if the groups in no way correspond with traditional racial divisions, and even if these DNA differences are quantitative not qualitative," that is, even if these are differences in the *frequency* of shared genetic traits rather than differences in the *types* of genetic traits. Under these circumstances the continued use of race and ethnicity in genetic research is unlikely to avoid the scientific and social pitfalls of the past.

Moreover, because geneticists have been able to co-opt principal components from the social constructionist critique of this practice, it will be necessary to rearticulate this critique if it is to succeed in preventing the (further) geneticization of race, ethnicity, and other forms of social identity. It seems likely that this will require a much better understanding of how geneticists use these as variables in their research — to identify instances where their use undermines science and is, therefore, counter-productive. In this way it may be possible to persuade geneticists that, even when they separate the use of race and ethnicity from the social world in which these are produced, the fundamental social nature of these concepts imposes critical limits on their utility as scientific variables.

ACKNOWLEDGMENTS

This study would not have been possible without the participation of the interviewees, all of whom are busy academics with other pressing priorities and all of whom generously made time to discuss these issues with us. We hope that our analyses have done justice to their contribution and faithfully represent the diversity of views and experiences they shared with us. The analyses we present and any mistakes these contain are our responsibility alone. We are also grateful to Thea de Wet, Richard Tutton, and Alan Goodman, who helped clarify the arguments we have made, and to the Wellcome Trust's research program in biomedical ethics, which supported this research (Grant 057182).

REFERENCES

Banton, M., *Racial Theories*, 2nd ed., Cambridge University Press, Cambridge, 1998.

Braun, L., Race, ethnicity, and health: can genetics explain disparities? *Perspectives in Biology and Medicine*, 45, 159–174, 2002.

Cashmore, E., Banton, M., Jennings, J., et al., *Dictionary of Race and Ethnic Relations*, 4th ed., Routledge, London, 1996.

Cooper, R.S. and Freeman, V.L., Limitations in the use of race in the study of disease causation, *Journal of the National Medical Association*, 91, 379–383, 1999.

Ellison, G.T.H., De Wet, T., IJsselmuiden, C., et al., Desegregating health statistics and health research in South Africa, *South African Medical Journal*, 86, 1257–1262, 1996.

Ellison, G.T.H., De Wet, T., IJsselmuiden, C., et al., Segregated health statistics perpetuate 'racial' stereotypes, *British Medical Journal*, 314, 1485, 1997.

Ellison, G.T.H. and Jones, I.R., Social identities and the 'new genetics': scientific and social consequences, *Critical Public Health*, 12, 265–282, 2002.

Friedman, D.J., Cohen, B.B., Averbach, A.R., et al., Race/ethnicity and OMB Directive 15: implications for state public health practice, *American Journal of Public Health*, 90, 1714–1719, 2000.

Gannett, L., Racism and human genome diversity research: the ethical limits of 'population thinking,' *Philosophy of Science*, 68, 479–492, 2001.

Gilbert, G.N. and Mulkay, M., *Opening Pandora's Box: A Sociological Analysis of Scientists' Discourse*, Cambridge University Press, Cambridge, 1984.

Glaser, B.G., *Grounded Theory Perspective: Conceptualization Contrasted with Description*, Sociology Press, Mill Valley, CA, 2001.

Goodman, A.H., Why genes don't count (for racial differences in health), *American Journal of Public Health*, 90, 1699–1702, 2000.

Krystufek, B., First record of the garden dormouse (*Eliomys quercinus*) in Slovenia, *Acta Zoologica Academiae Scientiarum Hungaricae*, 49, 77–84, 2003.

Lewontin, R.C., An apportionment of human diversity, in *Evolutionary Biology 6*, Dobzhansky, T., Hecht, M.K., and Steere, W.C., Eds., Appleton Century Crofts, New York, 1972, pp. 381–398.

Lippman, A., Prenatal genetic testing and genetic screening: constructing needs and reinforcing inequalities, *American Journal of Law and Medicine*, 17, 15–50, 1992.

Livingstone, F.B., On the nonexistence of human race, in: *Concept of Race*, Montague, A., Ed., The Free Press, New York, 1964, pp. 46–60.

McCann-Mortimer, P., Augoustinos, M., and Le Couteur, A., 'Race' and the Human Genome Project: constructions of scientific legitimacy, *Discourse and Society* 15, 409–432, 2004.

Nature Genetics, The unexamined 'Caucasian', Editorial, 36, 341, 2004.

Nazroo, J.Y., The structuring of ethnic inequalities in health: economic position, racial discrimination, and racism, *American Journal of Public Health*, 93: 277–284, 2003.

Nelkin, D. and Lindee, M.S., *The DNA Mystique: The Gene as Cultural Icon*. New York, Freeman Press, 1996.

NIH (National Institutes of Health), What is the National Institutes of Health (NIH)?, available at http://www.nih.gov/about/Faqs.htm#NIH, accessed January 2006.

Outram S.M. and Ellison, G.T.H., Anthropological insights into the use of race/ethnicity to explore genetic determinants of disparities in health, *Journal of Biosocial Science*, 2005, 38, 83–102.

Outram, S.M. and Ellison, G.T.H., Improving the use of race/ethnicity in genetic research: a survey of instructions to authors in genetics journals, *Science Editor*, 29, 2006.

Pfeffer, N., Theories of race, ethnicity, and culture, *British Medical Journal*, 317, 1381–1384, 1998.

Pinker, S., *The Blank Slate: The Modern Denial of Human Nature*, Penguin, London, 2003.

Relethford, J.H., Apportionment of global human genetic diversity based on craniometrics and skin color, *American Journal of Physical Anthropology*, 118, 393–398, 2002.

Root, R., The use of race in medicine as a proxy for genetic differences, *Philosophy of Science*, 70, 1173–1183, 2003.

Rosenberg, N.A., Pritchard, J.K., Weber, J.L., et al., Genetic structure of human populations, *Science* 298, 2381–2385, 2002.

Sankar, P., MEDLINE definitions of race and ethnicity and their application to genetic research, *Nature Genetics*; 34, 119, 2003.

Smaje, C., *Health, 'Race' and Ethnicity: Making Sense of the Evidence*, Kings Fund Institute, London, 1995.

Strauss, A.L. and Corbin, J.M., *Basics of Qualitative Research: Techniques and Procedures for Developing Grounded Theory*, Sage, London, 1998.

Tate, C. and Audette, D., Theory and research on 'race' as a natural kind variable in psychology, *Theory and Psychology*, 11, 495–520, 2001.

Tatum, B.D., *Why are All the Black Kids Sitting Together in the Cafeteria? and Other Conversations about Race*, Basic Books, New York, 1997.

Taussig, K.S., Rapp, R., and Heath, D., Flexible eugenics: technologies of the self in the age of genetics, in: *Genetic Nature/Culture: Anthropology and Science Beyond the Two-Culture Divide*, Goodman, A.H., Heath, D., and Lindee, M.S., Eds., University of California Press, Berkeley, 2003.

UNESCO, *Statement on Race*, UNESCO, Geneva, 1950.

UNESCO, *Statement on the Nature of Race and Race Differences*, UNESCO, Geneva, 1951.
UNESCO, *Universal Declaration on the Human Genome and Human Rights*, UNESCO, Geneva, 1997, available at http://portal.unesco.org/en/ev.php-URL_ID=13177&URL_DO=DO_TOPIC&URL_SECTION=201.html, accessed January 2006.
Wetherell, M., Taylor, S., Yates, S.J., Eds., *Discourse as Data: A Guide for Analysis*. Sage, London, 2001.
Witzig, R., The medicalization of race: scientific legitimization of a flawed social construct, *Annals of Internal Medicine*, 125, 675–679, 1996.
Wolf, P.G. and Soltis, P.S., Estimates of gene flow among populations, geographic races, and species in the *Ipomopsis-Aggregata* complex, *Genetics*, 130, 639–647, 1992.
Woolgar, S., Pawluch, D., Ontological gerrymandering: the anatomy of social problems explanations, *Social Problems*, 32, 214–227, 1985.

10

INDIGENOUS PEOPLES, BIOANTHROPOLOGICAL RESEARCH, AND ETHICS IN BRAZIL: ISSUES IN PARTICIPATION AND CONSENT

Ricardo Ventura Santos

CONTENTS

INTRODUCTION

Questions concerning the ethics of bioanthropological research among indigenous peoples have recently attracted a great deal of public attention. These questions have arisen both for events taking place in specific ethnographic and geographic contexts, such as those that led to the controversy surrounding Patrick Tierney's book *Darkness in Eldorado*, and those taking place on a global scale, like the much debated Human Genome Diversity Project (HGDP), which was originally intended to involve the collection of biological samples from hundreds of populations to capture genetic diversity across the world.* As the contributions to this volume make very clear, ethical concerns, which have generally been discussed as they pertain to research interests, now stand independently as favored themes for debates — debates promoting many different methodological and theoretical viewpoints and covering investigations of past peoples as well as those of the present.

The ethics of research, which are based on concepts such as autonomy, beneficence, nonmalfeasance, justice, and equity, are closely related to the defense and guarantee of human rights. A key document in the history of bioethics is the Nuremberg Code of 1947, the 10 points of which were drawn up against the backdrop of Nazi medical and scientific practices during World War II. The code is, therefore, a document produced in a specific context but one that advances principles intended to be applicable on a worldwide scale. Other international codes followed, such as the Helsinki Declaration, introduced in 1964, which has gone through many revisions over the years. These and related international documents are often the fundamental references used by countries to construct their own codes of ethics in research.

While the principles behind the various codes of ethics in research are practically the same in all Western countries, their regulation varies enormously from country to country. By *regulation* I mean the functioning of the diverse judicial and institutional agencies involved in the implementation of codes, laws, and directives related to ethics in research. This diversity in the forms of regulation, which results from historical, social, and political circumstances interacting with cultural factors, calls for an anthropological perspective. As pointed out by the physician and anthropologist Arthur Kleinman, in an essay entitled *Anthropology of Bioethics*,

> What characterises anthropological approaches to ethical issues, in medicine as well as other fields, is an emphasis on questions that emerge out of the grounded experiences of sick persons,

* For the controversy surrounding Patrick Tierney's book see AAA (2002), Salzano and Hurtado (2004), and Coronil et al. (2001). For the HGDP see Reardon (2001) and Santos (2002).

> families, and healers in concrete contexts . . .. In place of universalist and essentialist propositions — philosophical and political economic — anthropologists . . . have focused upon the interactions of everyday life, the societal hierarchies and inequalities they represent, and the moral issues in which they are clothed. Thereby, anthropologists examine ethics at the intersection of the social logics of symbolic systems, social structures, and historical events. (Kleinman, 1995: 45).

In this chapter I focus my analysis on the concepts of *participation* and *consent* as laid out in Brazilian government documents to demonstrate the impact of historical sociopolitical factors on the system that regulates ethics in research among indigenous peoples in Brazil. While the concepts invoked in these documents are in harmony with internationally accepted principles of research ethics (autonomy, beneficence, nonmalfeasance, justice, and equity), the way they are put into practice is shaped by historical precedents that condition and prescribe the interaction of the nation-state with indigenous peoples in Brazil. The elements Kleinman (1995) highlights (interactions of everyday life, societal hierarchies and inequalities, social logics of symbolic systems, social structures, and historical events) are key to understanding these relationships.

This analysis is driven by practical motivations and concerns. When I was recently reading documents related to the controversy generated by *Darkness in Eldorado*, I was struck by the lack of reference to regulations governing the ethics of research in Venezuela and Brazil, the countries where Yanomami lands are located. Even though the events discussed took place in South American countries, the principles regulating research ethics in those countries were hardly mentioned (see Chernela, 2002; Hurtado et al., 2001; Watkins, 2002). I should emphasize that it is not my intent to criticize what might be called a certain bioethical ethnocentrism present in these debates. What I wish to demonstrate, by calling attention to the regulation of research in a specific national setting (i.e., taking the Brazilian case as an example) is, among other points, that these regulations and the operationalization of ethics in research are the products of historically contingent sociopolitical processes. These processes reflect the interpretation of human biological diversity and determine how knowledge about human biology is generated (for example, by affecting the feasibility of carrying out certain types of investigation in biological anthropology).

THE ROOTS OF STATE GUARDIANSHIP OF INDIANS IN BRAZIL

In the eighteenth century, during a period of intense European expansion, long voyages, and contact with peoples from the far corners of the earth, the nature of the differences among human beings was hotly debated. Eighteenth-century philosophers like Rousseau defended ideas of human unity and believed that all humans had equal capacity for improvement (the gift of "perfectibility"). At the same time that some held this vision, other lines of argument emphasized the existence of deep and primordial differences among human beings. The natives of the American continents were an important focus of this debate (Gerbi, 1996; Jahoda, 1999; Schwarcz, 1993; Todorov, 1983).

In this context of European contact with new worlds, new plants, new animals, and new human beings, thinkers of the stature of Buffon, the famous French naturalist, developed the thesis concerning the "debility" and "immaturity" of the peoples of the Americas. To Buffon, nature on the American continent, when compared to that in the Old World, appeared to be less active, less varied, and more fragile. Buffon, who based his arguments mainly on the characteristics of the flora and fauna, found an important example in the quadrupeds. The puma, in Buffon's view the equivalent of the lion, has no mane; "also he is much smaller, weaker, and more cowardly than the true lion" (see Gerbi, 1996:19; Jahoda, 1999:19–23). The weakness of American nature was confirmed by the fate of domestic animals introduced by Europeans. It was thought that with the passage of time they became atrophied, shrank in size, and were reduced to dwarfs or miniature caricatures of their prototypes.

This weakness and immaturity, according to Buffon, also characterized the natives of the American continents, whom he presented as childlike creatures:

> For although the savage of the New World is of almost the same stature as the men of our world, that does not suffice for him to be an exception to the general rule of the reduction of living nature in the whole continent. The savage is feeble and small in his organs of generation; he has neither body hair nor beard, and no ardour for the female of his kind. Although lighter than the European, on account of his running more, he is nevertheless much less strong in body; he is also much less sensitive, and yet more fearful and more cowardly; he lacks vivacity, and is lifeless in his soul; the activity of his body is less an exercise or voluntary movement than an automatic reaction to his needs; take from him hunger and thirst, and you will destroy at the same time the active cause of all his movements; he will remain either standing there stupidly or recumbent for days at a time." (cited in Gerbi, 1996:21; Jahoda, 1999:20)

Buffon's ideas were conceived in a European intellectual climate that was rapidly moving toward models that explained human diversity in terms of the concept of race. As we know, these models reached their apogee between the mid–nineteenth century and the beginning of the twentieth century and were considerably influenced by the consolidation of evolutionism in European and American science and society (Gould, 1996; Stocking, 1968; Todorov, 1983).

The idea that Amerindians are immature, fragile, and childlike creatures has a complex, multifaceted, and surprisingly long history. In Brazil it remained influential well into the twentieth century.* In the 1910s Edgard Roquette-Pinto, the most prominent physical anthropologist in Brazil at the time, who worked at the National Museum in Rio de Janeiro, participated as a naturalist in an expedition that crossed immense stretches of the then unexplored interior of Brazil (Figure 10.1). Members of the expedition, led by the famous Brazilian explorer Candido Rondon (Figure 10.2), whose main objective was to install telegraph lines across the interior of the country, contacted many previously isolated Indian groups such as the Nambikwára, a group that, a few decades later, would be described by Claude Levi Strauss in his classic ethnography, *Tristes Tropiques*. Roquette-Pinto wrote down some thoughts about the Indians in his book, *Rondonia: Antropologia – Etnografia*:

> It should not be our concern to make them citizens of Brazil. Everyone understands that an Indian is an Indian, a Brazilian is a Brazilian. The nation should protect them and even support them, just as it willingly accepts the burden of supporting abandoned or indigent children, the sick, and the insane. Abandoned children, and even the insane, work, but society does not support them in order to take advantage of their efforts . . .. The program should be to protect them [the Indians], but not to direct them, so that their spontaneous evolution is not compromised." (Roquette-Pinto, 1917:200–201)

In Brazil this line of reasoning, charged with the positivism of the French philosopher August Compte (which equated the Indians to children and other social segments in need of protection), was not confined to a small group of

* Alcida Ramos (1998) points out that the view of indigenes (and generally of foreign peoples) as being like children goes back a very long way in Western thought, at least as far as Aristotle. In the case of the New World, the image of the childlike native appeared in the writings of the Portuguese and Spanish chroniclers at the time of the discoveries, and it continued on through later centuries.

Figure 10.1 A Nambikwára man photographed by Roquette-Pinto in 1912 (from Roquette-Pinto, 1917:254–255).

intellectuals. It was also expressed in the Brazilian Civil Code of 1916 as the notion of "legal competence." This code affirmed that "every man is capable of his civic rights and duties." However, it distinguished two levels of "incompetence": those who are "absolutely incompetent" (minors under

Figure 10.2 Cândido Rondon distributing gifts to the recently contacted Aritī from Mato Grosso (photo by Major Thomaz Reis, reproduced in Rondon, 1946:85).

the age of 16, "the insane of all kinds," and deaf-mutes, "who cannot express their will") and those "incompetent relative to certain acts" (over the age of 16 but under the age of 21), "prodigals,"* and "forest-dwellers." It was considered that the Indians, while not totally incapable of exercising their rights, should remain "under guardianship" until they became integrated into "the civilization of the country."**

The image of Indians as almost children was extremely influential in shaping the Indian policy in effect in Brazil through most of the twentieth century, as administered by government agencies such as the Serviço de Proteção aos Índios (Indian Protection Service; SPI) and the Fundação Nacional do Índio (National Indian Foundation; FUNAI: see Cunha, 1987; Freire, 1996; Lima, 1992, 1995; Ramos, 1998). As Farage and Cunha (1987) point out:

> . . . in the beginning of the twentieth century, indigenous societies appeared as infantile forms . . . which should be guided by means of guardianship toward the civilization of our society. Guardianship, which above all was to have been a state instrument to defend indigenous lands, was then discussed in totally inadequate terms . . . that took for granted the infantile character of Indians and their societies. (1987:114; cited in Ramos, 1998:18)

Cunha (1987) shows that the concept of guardianship, while discriminatory in a certain sense, was intended to serve as additional protection for the Indians, although it did not always work that way in Brazilian indigenous policy. It was intended as protection, permitting the Indians to live according to their own social norms (since they had "imperfect" knowledge of the surrounding society) and providing them with special attention so they would not be harmed or defrauded. However, rather than "a special zeal in favour of those it was charged to protect" (Cunha, 1987:28), in practice guardianship often became a form of control and coercion, with the interests of the Indians taking second place.

---

* "Prodigals," as in the prodigal son, are persons who adopt "irresponsible behaviour."

** In the view of some jurists, no great distance separated Indians from other social segments classified as "incompetent." Carlos Frederico Marés de Souza Filho (1994:162) calls attention to the comments of a well-known Brazilian jurist, Nelson Hungria, who, writing in the 1950s, observed, "Article 22 [of the penal code] speaks of incomplete or retarded development. Under this rubric fall not only those who are congenitally deficient in mental development, or feeble minded (idiots, imbeciles, morons) but also those who lack certain senses (deaf-mutes) and even in adapted forest-dwellers . . . thus there is no doubt that among the mentally deficient we may also include *Homo sylvester*, as totally lacking the ethical attainments of civilized *Homo medius* who is deemed responsible under the penal code."

Up until the present day, guardianship continues to govern relations between the Indians and the Brazilian State.* For example, the principal law in effect in Brazil that concerns indigenous peoples is the Estatuto do Índio (Indian Statute) passed in 1973. According to this statute Indians, are considered *a priori* to be without "competence" from a civil point of view; they can gain their civil rights only when freed from tutelage. Another striking aspect of the Indian Statute is the heavy emphasis on assimilationism, which is clearly expressed in Article 1: "[this law] regulates the juridical situation of the Indians or forest-dwellers and of indigenous communities, with the aim of preserving their culture and of integrating them, progressively and harmoniously, into the national communion."**

A significant development, which contributed to modifying the emphasis on assimilationism, was the promulgation of the Federal Constitution of 1988. According to Souza-Filho (1994), the new Constitution transformed relations between the state and indigenous peoples, implementing ideas that broke with five centuries of integrationist policies. The Constitution establishes the principle that the Brazilian state should protect Indian rights, but it does not mention guardianship, Indian agencies, or the civil incompetence of Indians. The Constitution significantly increased the rights of Indians, recognizing their social organizations, their practices, their religions, their languages, and their cultures. According to Souza-Filho (1994:218–219), "above all, it called the Indians 'Indians' and gave them the right to continue as such."

As a result of these changes, the body of principles and laws in effect in Brazil is marked with serious internal discrepancies as it applies to indigenous peoples. The Indian Statute, conservative and assimilationist, continues to exist within a wider judicial system, represented by the Constitution of 1988, which invests the Indians with more extensive rights of citizenship.***

* Although they are a very small minority in demographic terms, indigenous peoples have a powerful presence in the historical, social, and cultural imagination of the Brazilian people. The anthropologist Alcida Ramos (1998:3) asked recently why "being so few, [Indians in Brazil] have such a prominent place in the national consciousness?" She herself answered, "They have the power to burrow deeply into the country's imagination."

** The reformulation of the Indian Statute has recently been under discussion. One of the new proposals was to revoke the guardianship power, substituting it with other means of protecting the Indians' collective rights (ISA, 2002a,b).

*** The Indian Statute does provide the possibility of emancipation. Individual Indians or indigenous communities may petition for emancipation. Requirements are "a minimum age of 21, knowledge of the Portuguese language, ability to exercise a useful activity, and an understanding of the uses and customs of the national communion" (Article 9, Indian Statute).

THE SYSTEM REGULATING BIOETHICS IN BRAZIL

This brief historical account is essential to the analysis of the most important body of directives and standards that regulates research on human subjects in Brazil, the *Guidelines and Norms Regulating Research Involving Human Subjects* (also known as Resolution 196/1996), which was confirmed by the Ministry of Health in 1996 (Conselho Nacional de Saúde, 2002). Resolution 196/1996 is based on the principal international documents that set forth norms of ethics in research involving human subjects: the Nuremberg Code (1947), the Declaration of Human Rights (1948), the Helsinki Declaration (1964 and later versions), as well as other international treaties and Brazilian legislative decisions. It incorporates, from both individual and collective viewpoints, the principal references for bioethics, seeking to assure the rights and duties of the scientific community, the subjects of research, and the state. It stipulates the necessity of: (a) obtain free and informed consent of individuals who are the subjects of research and protecting vulnerable groups and the legally incompetent (*autonomy*); (b) weighing benefits against risks, both present and potential, individual and collective (*beneficence*), with a commitment to seek maximum benefits and minimum risks and damages; (c) guaranteeing that foreseeable harm will be avoided (*nonmalfeasance*); and (d) conducting research that is socially relevant, with significant advantages for its subjects and minimal demands on vulnerable participants, which guarantees equal consideration for the interests involved without losing the sense of its social and humanitarian purpose (*justice and equity*).

In applying Resolution 196/1996, some groups are differentiated in terms of the procedure for obtaining free and informed consent. The resolution indicates that "in cases where there is some restriction to freedom or to the capacity of understanding needed for adequate consent" (those involving children and adolescents, the mentally disturbed or ill, and subjects substantially diminished in their capacity for consent), consent should be obtained through the legal representatives of the subjects, maintaining the right of the individual to be informed. Significantly, in the resolution, Indians do not appear as one of the groups characterized by "restriction to the freedom or understanding necessary for adequate consent" with which they were formerly classified in the Brazilian judicial system.

Nevertheless, Indians do have a differentiated space in Resolution 196/1996, as they appear under what are referred to as "special thematic areas." Indigenous populations are listed under this category, along with human genetics; human reproduction; pharmaceuticals; medicines; vaccines and new diagnostic tests; new health equipment, supplies, and appliances; procedures not yet well established in the literature; projects that involve biosafety; research coordinated outside the country or with foreign participation; and research that involves sending biological materials to foreign countries. Notably, among the various "special thematic areas" only the one that refers to Indians is defined by sociological aspects or by population boundaries, while the others generally refer to aspects of research and technology.

The designation of "special" conferred on indigenous populations in Resolution 196/1996 is hardly trivial, as it is part of a complex network of historical, political, and sociological factors, some of which I have already mentioned. Indians are no longer equated with "children and adolescents, those mentally disturbed or ill, and subjects substantially diminished in their capacity for consent," but they are still placed apart. One might ask what the connection is between the attribution of "special" in Resolution 196/1996 and the idea of guardianship that is so central to Brazilian indigenous policy and that is maintained in effect by the Indian Statute. On the one hand, the notion of "relative incompetence," the explicit basis of guardianship policy, is not found in Resolution 196/1996. In this resolution the "special" status of indigenous peoples is coupled with the need to respect their social and cultural differences, without reference to assimilationist ideas. The Resolution also emphasizes direct participation of the Indians in deciding whether or not the proposed investigation will be carried out. On the other hand, the attribution of "special" carries with it the requirement that research projects involving indigenous populations be analyzed by a central agency, the Comissão Nacional de Ética em Pesquisa (National Commission for Ethics in Research; CONEP), seated in Brasilia, which is responsible for "the protocols for research in special thematic areas." Therefore, it could be argued that sustained vigilance over indigenous societies by the Brazilian state is still evident in ways that are inherent to and, to some extent, reminiscent of the tradition of control and guardianship described earlier.*

HEALTH AND THE REGULATION OF ETHICS IN RESEARCH AMONG INDIGENOUS PEOPLES

I have tried to show that the regulations governing ethics in research concerning indigenous peoples in Brazil are deeply embedded in the history of their relationship with the Brazilian state. In this section I shall explore a different side of the question, that is, the growing prominence

* As well as being approved by the committees on ethics, research projects involving indigenous populations must also be submitted to the Fundação Nacional do Índio (National Indian Foundation; FUNAI), whose directive 0475/1988 sets up "norms for the concession of permission for outsiders to enter indigenous reservations for the purpose of carrying out scientific research." Thus, this process requires going through many government agencies — something that might be interpreted as extra vigilance by the state.

of a specific aspect of regulation. At a time when public policies in Brazil are being transformed, medical care and health are becoming an important setting for decision-making in the system regulating ethics in research.

Until recently, the services providing health care to indigenous populations in Brazil were under the management of the Fundação Nacional do Índio (National Indian Foundation; FUNAI), which provided mainly curative medical care. There was little continuity in providing basic preventive health care to indigenous communities. Even in the 1960s and 1970s, when vaccines, antibiotics, and other resources were available to deal with a number of diseases, the government services provided by FUNAI failed to prevent epidemics of tuberculosis and measles from killing hundreds of Indians recently contacted in Central Brazil and the Amazon. This happened especially in the Amazon, where economic frontiers were advancing. Even in the absence of serious epidemics, with few exceptions, the services managed by FUNAI tended to be disorganized and, in many areas, sporadic (Coimbra et al., 2002; Coimbra and Santos, 2004).

In the 1980s and 1990s FUNAI went through a period of great instability, with frequent political and structural changes, that had a negative effect on its status in the eyes of the Indians, on the morale of its functionaries, and also on how it was regarded more generally by public opinion. FUNAI seemed to be drifting, and the successive management changes produced no clear directives to make FUNAI's work with the Indians any more effective in the sphere of education or health care or in demarcating reservation lands. Health care for the Indians was particularly disorganized and sporadic during this period, with little improvement in the most elementary of health measures, either curative or preventive (Coimbra et al., 2002, Coimbra and Santos, 2004).

In the 1980s the Brazilian health sector went through a substantive transformation, which ended with the creation of the Sistema Único de Saúde (Unified Health System; SUS), inspired by the Federal Constitution of 1988. Basically, SUS stresses the role of the state in financing, organizing, and implementing a universal and egalitarian health care system. As a by-product of the expansion of SUS, since 1999 the Fundação Nacional de Saúde (National Health Foundation; FUNASA), a branch of the federal Ministry of Health, has held the responsibility for providing health care to indigenous peoples, replacing FUNAI. It set up health services designed especially for indigenous peoples, and districts were divided according to ethnic and geographical criteria (the so-called special indigenous health districts; DSEI). Throughout the country, 34 districts are now in place. Each district covers a territory that does not necessarily coincide with municipal boundaries but is defined in technical and ethnodemographic terms (Coimbra and Santos, 2004; FUNASA, 2002).

In terms of organization, the DSEIs work through district and local "councils." The councils are permanent deliberative associations, made up of government representatives, service providers, health professionals, and service users. They devise strategies and control the execution of health policies, including economic and financial aspects.* In the case of indigenous peoples, the councils must engage the participation of community representatives, including traditional leaders, indigenous teachers, indigenous health workers, traditional health specialists, and birth attendants. The entire process of setting up health services for indigenous peoples is firmly grounded in the participation of the Indians themselves through their organizations and leadership (FUNASA, 2002).

The direction taken by public policy for health care has influenced the regulation of ethics for research among indigenous peoples. Resolution 196/1996 indicated that specific norms should guide the ethics of research in so-called special thematic areas. In 2000, Resolution 304/2000 was passed, which takes up specific questions related to research among indigenous peoples. This resolution not only requires that approval for research projects pass the direct scrutiny of the communities; it also states that the benefits and advantages of the investigation "must serve the needs of the individuals or groups that are the subjects of study, or of similar societies, and/or the national society, taking into consideration the promotion and maintenance of well-being, the conservation and protection of biological and cultural diversity, individual and collective health, and contribution to the development of the community's own knowledge and technology." There is, therefore, an emphasis on research that will provide an immediate return to the community.

RESEARCH ETHICS IN BRAZIL AND THE DIRECTION OF RESEARCH IN BIOLOGICAL ANTHROPOLOGY

Since 1990, the research group to which I belong has been carrying out research in public health, human ecology, and human biocultural variation among the Xavánte, an indigenous people of Mato Grosso, in Central Brazil (see Coimbra et al., 2002). During a recent visit to the Etéñitépa community, where we did most of our field research, we had the opportunity to hear the thoughtful words of an influential leader, Tsuptó Bupréwen Wairi. His comments touch on important questions related to scientific research, to investigators, and to the participation of indigenous communities:

* The organization of the health councils is defined in federal law no. 8.142/1990, which regulates the participation of the community in the administration of the SUS, as well as intergovernmental transfers of financial resources in the area of health.

> In former times . . . there were not so many sicknesses . . . there wasn't the interference that there is today, the interference of the Whites. At that time there was no tuberculosis, no diabetes . . . Now we have them all. These sicknesses are not from our village; they are not from other Indian populations . . . Our organisms, our bodies, the bodies of our children, can't stand up to all this; they can't resist . . . About the work of research, I think the research team has to relate to the community; the team has to like what it is doing . . . I think that what is affecting our health, what is damaging it, these things that come into the village from outside, have to be investigated . . .. The research that has already been done has helped a lot, even if it is over the long term. But we see this as an example of how we can find out what is damaging the health of the village, what may damage it in the future . . . We also see research as a way for people outside to learn about our lives. We have health problems that people outside don't know about, so they don't look for solutions . . . Well, I think these things should be studied in more depth so that if solutions for these things that are happening can be found, they will be. (cited in Coimbra et al., 2002:xix–xx)

Tsuptó's words are revealing on several levels. They acknowledge the benefits of research, especially research on topics important to the community that may help make assistance programs more effective. They also show his awareness that research does not necessarily solve problems in the short term but may require sustained work over a long period to produce results. In the past few years a number of projects have been developed at Etéñitépa, including, among others, the recovery and maintenance of biodiversity in the *cerrado* environment, audio and video taping of cultural practices, and the implementation of health care projects. A certain amount of research was associated with each of these projects (see Coimbra et al., 2002; Graham, 2000).

A significant feature of the Xavánte leader's testimony is his comment on the importance of research that documents the pattern of inequality between indigenous peoples and the surrounding national society. When he states that "we also see research as a way for people outside to learn about our lives" and at the same time points out that ". . . the diseases that the Indian populations are suffering from are brought by the Whites," Tsuptó is giving a clear signal that scientific research can and should lead to community empowerment.

The Xavánte leader's words also reveal to what degree research among indigenous peoples in Brazil has become politicized. They hint at the many expectations, not always apparent at first sight, that are at stake. Tsuptó is the leader of a community that has broad connections, with both governmental and nongovernmental organizations, and he has traveled widely in Brazil and abroad. While his leadership style certainly has special features, it may not be exceptional within the present context of ethnic politics in Brazil. Tsuptó's views on scientific research are essentially quite close to those that are emerging in forums like the Third National Conference on Indigenous Health, a meeting held in Luziânia (Goiás) in 2001 that brought governmental and nongovernmental organizations together with hundreds of indigenous representatives. In a document that was approved at this meeting, *Ethics in Research, Intellectual Property, and Patents Involving Indigenous Peoples*, a major point was that "research projects should always be designed to help communities solve their problems in seeking to improve their quality of life" (Anonymous, 2001). This emphasis on applied research, seen in Brazilian government documents as well as in indigenous movement manifestos, has important repercussions for research in biological anthropology among indigenous peoples in Brazil.

In this climate, it is likely to be easiest to obtain consent and community participation for research that has the direct potential to answer applied questions or is directly related to health (such as the evaluation of health needs and health services). The way ethics in research is regulated, with heavy reliance on the health system (through the role played by the district and local councils), strongly favors this trend. Among the research topics that interest biological and medical anthropologists, the socioenvironmental and epidemiological determinants of malnutrition and infectious and parasitic diseases stand out, as they are fundamental aspects of the health and disease profile of indigenous peoples in Brazil. These are topics that the research community, health professionals, and indigenous communities recognize as having high priority.

However, the immediate future of other fields of biological anthropology is harder to ascertain, because of debates about research ethics and priorities. Among these are investigations of human genetic diversity. At present, for various reasons, indigenous communities in Brazil are exhibiting a great deal of resistance to this kind of research (see Ramos, 2000; Santos, 2002). Not only are there no legal regulations to control how royalties would be shared for research leading to commercial products, but past events have increased suspicion about the motives of research into human genetic diversity. A significant example is the controversy surrounding biological specimens from two indigenous groups of Brazilian Amazonia (the Karitiána and the Suruí of the state of Rondônia), which has received wide attention in Brazil since 1996 (Santos, 2002). These samples are included in the so-called Human Diversity Collection and are deposited in the Coriell Cell Repositories, in Camden, New Jersey. Leaders of both indigenous groups, with the support of nongovernmental organizations, have opposed the retention of the samples in the repository, questioning the existence of permission to collect them and protesting against the wider availability for research of biological samples obtained without specific approval for that purpose. The disposition of the samples has become a major sticking point in discussions concerning research on human genetic variation among indigenous peoples in Brazil. Largely because of the debates over the Karitiána and Suruí specimens, a paragraph was included in Resolution 304/2000 stipulating that no DNA banks, cell lines, or other biological materials may be collected from indigenous peoples without the express consent of the communities involved.

Not only in Brazil, but also in other parts of the world, a serious challenge facing biological anthropologists and other scientists who study human genetic variation is to encourage the participation of communities while protecting them against possible injury, including improper commercialization of biological materials (DeCamp and Sugarman, 2004; Greely, 2001a; Sharp and Foster, 2000; Weijer et al., 1999). Long-term storage of biological specimens and the possibility of their commercialization are not the only issues. Chernela points out, in the context of the debate over what should be done with the Yanomami samples collected by James Neel and his collaborators in the 1960s, that Davi Kapenawa and other Yanomami leaders favor return of the material because, among other reasons, they "may not assign positive value to reconstructions of their past" (Chernela, 2002) through genetic research.

If anything has become clear in the present climate of confusion and suspicion that surrounds research in human genetic variation of indigenous peoples in Brazil and in other Latin American countries, it is the necessity to put a great deal of effort into negotiation and compromise in the near future.* Beyond the points already mentioned, it will be essential to discuss how the complex category of "research projects designed to help communities solve their problems in seeking to improve their quality of life" (Anonymous, 2001) may apply to research on human genetic variation.

In recent years a range of ideas has been brought to bear on certain aspects of bioethics, seeking to make them more relevant to the real conditions of indigenous peoples, with an emphasis on consent and participation in the area of population genetics (see Greely, 2001a, 2001b; Foster et al., 1997, 1998). Topics under discussion include: the definition of the scope of research; informed consent; the distribution of royalties from intellectual property; the future use of archival samples; and specific cultural concerns. This has led to an emphasis on the identification of culturally appropriate public and private social structures within which community members are accustomed to make decisions about health, aiming at the generation of models useful for reaching a communal consensus about genetic research (Figure 10.3). These are important points

* For a general discussion on this topic, see DeCamp and Sugarman, 2004.

Conduct survey of individual health-care decision-making

⇓

Identify culturally-appropriate decision making units (public and private)

⇓

Consult community members about inclusion of representative units

⇓

Assist in formation of Community Review Board

⇓

Initiate dialogue between researchers and community in public meetings

⇓

Internal discourse in public and private social units

⇓

Negotiations between researchers and Community Review Board

⇓

Community consensus and agreement

⇓

Researchers' intitutional review board evaluates documentation of communal discourse

⇓

Recruitment of subjects from community

⇓

Standard informed consent for individuals

Figure 10.3 General model put forward by Foster et al. (1998) for reaching communal consensus about genetic research.

of departure, and they are directly applicable to the current situation of bioanthropological research among indigenous peoples in Brazil.*

* Elsewhere I have addressed other important issues concerning ethics in research pertaining to indigenous peoples in Brazil, such as obtaining informed consent (Coimbra and Santos, 1996). As pointed out by several anthropologists, this is a highly complex issue. Foster et al. (1998:697) called attention to the dangers of attempting to apply to indigenous societies "the highly individualistic decision-making process idealized in Euro-American culture and enshrined in Western medical ethics." The essays published in the volume recently edited by Victora et al. (2004) bring together the critical perspective of a group of Brazilian anthropologists on this issue.

CONCLUDING REMARKS

Considering the social diversity that exists in Brazil, very little is known about its indigenous populations, whatever the dimension that might be under analysis. Today over 200 groups, speaking approximately 180 languages, are distributed throughout practically the entire national territory, in rural and urban areas, with great diversity in social and cultural characteristics, as well as historical, economic, and political trajectories. These range from a few small and still relatively isolated groups in the Amazon to others with substantial proportions of their populations dispersed in cities throughout the country. Indeed, it is even unclear precisely how many indigenous people there are because, depending on the source, they add up to anywhere between 350,000 and 700,000 people, but never surpassing 0.5% of the national population.

In recent years, it has become increasingly evident that there is a need to develop studies and establish strategies for gathering and analyzing data on indigenous peoples to better understand their realities and thereby plan and implement public policies in key areas of social policy such as education and health. It is not difficult to demonstrate how the invisibility of indigenous peoples in the national Brazilian databases can be negative, as can be seen clearly in the field of health. As pointed out by Coimbra and Santos (2004), it is not possible to satisfactorily sketch the health profile of indigenous peoples from the data currently available because these data lack the quantitative elements necessary for comprehensive analyses (see also Santos and Coimbra, 2003). In general, it is difficult to do more than compile studies of specific cases. Nevertheless, there is little doubt that the health conditions of indigenous peoples in Brazil are precarious, placing them at a disadvantage relative to other segments of the national society. Corroborating this diagnosis, a recent document from the Ministry of Health, *The National Politics of Healthcare of Indigenous Populations*, not only exposes the absence of data, but also suggests the magnitude of the inequalities between the health of indigenous populations and other segments of Brazilian national society:

> Unavailable are reliable global data regarding the health situation . . . [of indigenous populations], while those that are available are biased . . .. However precarious, available data indicate, in diverse situations, rates of morbidity and mortality three to four times greater than those encountered in the general Brazilian population. The large number of deaths that are unregistered or indexed without definite causes confirm the data's poor coverage and low capacity for resolving available services. (FUNASA, 2002:10)

In this important discussion about the invisibility of indigenous peoples and the need for more precise data, research activities occupy a very important role. It is important to point out that the politics of health care for indigenous populations in Brazil requires that such research should happen "in a differentiated manner, according to the cultural, epidemiological, and operational specifics of these peoples" (FUNASA, 2002:6). It is unnecessary to emphasize the relevance that biological anthropology and related fields can play in this agenda to support a better understanding of reality and thereby improve the living conditions and health of the indigenous populations of Brazil (see Coimbra et al., 2002; Coimbra and Santos, 2004; Santos and Coimbra, 1998, 2003). In many instances, it has been the indigenous movement itself that has demanded collaboration with researchers in order to characterize problems and develop appropriate interventions.

We are living in a unique time, characterized by a delicate equilibrium. At the same time that there is a consensus regarding the need for more efforts to understand local realities (and, therefore, to expand research activities) to support the efficient implementation of public policies, and to respond to the demands of communities, the requirements of the authorities controlling research ethics are becoming ever more prominent. I am not suggesting that research ethics should be looked on as an obstacle. However, in practice there are ever greater demands for conducting research, given that the steps for obtaining permissions have become so complex. It is worth emphasizing that, these days, the regulatory mechanisms for research ethics in Brazil with respect to indigenous populations give these populations an active and prominent voice in the process, which is very positive and helps reverse decades of paternalism.

In this context, I consider it relevant to point out that researchers in the field of biological anthropology and others need to be ever more resourceful in order to navigate the increasingly complex ethical regulatory environment for research. Accordingly, understanding the historical and sociopolitical genesis of regulation can be instructive. As I have sought to demonstrate in this chapter, in the case of the indigenous peoples of Brazil, although the precepts that regulate research ethics may be those with which the scientific community is familiar (such as autonomy, beneficence, nonmalfeasance, justice, and equity), there is a notable specificity in the mechanisms of regulatory functioning. These regulations have deeply embedded historical, social, political, and cultural elements that have grown out of the interrelationships of ethnic minorities and the Brazilian nation-state. I have also argued that, at the present time, certain areas of knowledge (including some that are part and parcel of biological anthropology) are likely to enjoy a more favorable reception, thereby influencing the chances that indigenous communities will accept and participate in such research activities.

Whatever the research goals, avoiding potential harm to research participants, maximizing benefits to them, ensuring a just distribution of the costs and benefits of research, and promoting trustworthiness of the research endeavor are basic principles that, above all, should govern bioanthropological research on and with the indigenous peoples of Brazil and beyond.

ACKNOWLEDGMENTS

I thank Nancy Flowers and James Welch for translating this text from Portuguese. I also thank George Ellison and Alan Goodman for inviting me to the conference where this paper was originally presented, as well as for their editorial comments. Carlos Coimbra Jr. read a preliminary version of this paper and made many helpful comments.

REFERENCES

American Anthropological Association (AAA), *Papers of the American Anthropological Association El Dorado Task Force*, American Anthropological Association, Washington, DC, 2002, available at http://www.aaanet.org/edtf/index.htm, accessed January 25, 2006.

Anonymous, *III Conferência Nacional de Saúde Indígena: Relatório Final*. Luziânia, Goiás, Brazil, 2001, available at http://www.rebidia.org.br/indigena/relafinal.htm, accessed January 25, 2006.

Chernela, J., Collection of bodily samples and informed consent: a discussion with recommendations, in *Papers of the American Anthropological Association El Dorado Task Force*, American Anthropological Association, Ed., Washington, DC, American Anthropological Association, 2002, available at http://www.aaanet.org/edtf/index.htm, accessed January 25, 2006.

Coimbra, C.E.A., Flowers, N.M., Salzano, F.M., et al., *The Xavánte in Transition: Health, Ecology and Bioanthropology in Central Brazil*, University of Michigan Press, Ann Arbor, 2002.

Coimbra, C.E.A. and Santos, R.V., Ética e pesquisa biomédica em sociedades indígenas no Brasil, *Cadernos de Saúde Pública*, 12(3), 417–422, 1996.

Coimbra, C.E.A. and Santos, R.V., Emerging health needs and epidemiological research in indigenous peoples in Brazil, in *Lost Paradises and the Ethics of Research and Publication*, Salzano, F.M. and Hurtado, A.M., Eds., Oxford University Press, Oxford, 2004, pp. 89–109.

Conselho Nacional de Saúde, *Resolução 196/1996: Diretrizes e Normas Regulamentadoras de Pesquisas Envolvendo Seres Humanos (Resolution 196/96: Guidelines and Norms Regulating Research Involving Human Subjects)*, Conselho Nacional de Saúde, Brasília, 2002, available at http://conselho.saude.gov.br/biblioteca/livros/Normas_Pesquisa.pdf, accessed January 25, 2006.

Coronil, F., Fix, A.G., Pels, P., et al., Forum on Anthropology in Public: Perspectives on Tierney's Darkness in El Dorado, *Current Anthropology*, 42, 265–276, 2001.

Cunha, M.C., Os índios no direito brasileiro hoje, in *Os Direitos dos Índios: Ensaios e Documentos*, Cunha, M.C., Ed., Editora Brasiliense, São Paulo, 1987, pp. 19–51.

DeCamp, M. and Sugarman, J., Ethics in population-based genetic research, *Accountability in Research*, 11, 1–26, 2004.

Farage, N. and Cunha, M.C., Caráter da tutela dos índios: origens e metamorfoses, in *Os Direitos dos Índios: Ensaios e Documentos*, Cunha, M.C. Ed., Editora Brasiliense, São Paulo, 1987, pp. 103–117.

Foster, M.W., Bernsten, D., and Carter, T.H., A model agreement for genetic research in socially identifiable populations, *American Journal of Human Genetics*, 63, 696–702, 1998.

Foster, M.W., Eisenbraun, A.J., and Carter, T.H., Communal discourse as a supplement to informed consent for genetic research, *Nature Genetics*, 17, 277–279, 1997.

Freire, C.A.R., A criação do Conselho Nacional de Proteção aos Índios e o indigenismo interamericano (1939-1955), *Boletim do Museu do Indio*, 5, 1996.

FUNASA (Fundação Nacional de Saúde), *Política Nacional de Atenção à Saúde dos Povos Indígenas*, Fundação Nacional de Saúde, Ministério da Saúde, Brasília, 2002, available at http://www.funasa.gov.br, accessed January 25, 2006.

Gerbi, A., *O Novo Mundo: História de uma Polêmica (1750-1900)*, Companhia das Letras, São Paulo, 1996. (In English: *The Dispute of the New World*, University of Pittsburgh Press, Pittsburgh, 1973).

Gould, S.J., *The Mismeasure of Man*, WW Norton & Company, New York, 1996.

Graham, L.R., Lessons in collaboration: the Xavante/WWF Wildlife Management Project in Central Brazil, in *Indigenous Peoples and Conservation Organization: Experiences in Collaboration,* Weber, R., Butler, J., and Larson, P., Eds., World Wildlife Fund, Washington, DC, 2000, pp. 47–71.

Greely, H.T., Human genomics research: new challenges for research ethics, *Perspectives in Biology and Medicine*, 44, 221–229, 2001a.

Greely, H.T., Informed consent and other ethical issues in human population genetics, *Annual Review of Genetics*, 35, 785–800, 2001b.

Hurtado, M., Hill, K., Kaplan, H., and Lancaster, J., The epidemiology of infectious diseases among South American Indians: a call for guidelines for ethical research, *Current Anthropology*, 42, 425–432, 2001.

Instituto Socioambiental (ISA), Os Índios Não São Incapazes, 2002a, available at http://www.socioambiental.org, accessed January 25, 2006.

Instituto Socioambiental (ISA), Estatuto do Indio, 2002b, available at http://www.socioambiental.org, , accessed January 25, 2006.

Jahoda, G., *Images of Savages: Ancient Roots of Modern Prejudice in Western Culture*, Routledge, London, 1999.

Kleinman, A., *Writing at the Margin: Discourse between Anthropology and Medicine*, University of California Press, Berkeley, 1995.

Lima, A.C.S., O governo dos índios sob a gestão do SPI, in *História dos Indios no Brasil*, Cunha, M.C., Ed., Companhia das Letras, São Paulo, 1992, pp. 155–172.

Lima, A.C.S., *Um Grande Cerco de Paz: Poder Tutelar, Indianidade e Formação do Estado no Brasil*, Editora Vozes, Petrópolis, 1995.

Ramos, A.R., *Indigenism: Ethnic Politics in Brazil*, University of Wisconsin Press, Madison, 1998.

Ramos, A.R., *The Commodification of the Indian*, *Série Antropologia*, no. 281, Departamento de Antropologia, Instituto de Ciências Sociais, Universidade de Brasília, 2000.

Reardon, J., The Human Genome Diversity Project: a case study in coproduction, *Social Studies of Science,* 31, 357–388, 2001.

Rondon, C.M.S., *Índios do Brasil do Centro, Noroeste e Sul de Mato Grosso*, Conselho Nacional de Proteção aos Índios, Ministério da Agricultura, Rio de Janeiro, 1946.

Roquette-Pinto, E., Rondonia (Anthropologia - Ethnographia), *Archivos do Museu Nacional*, 20, 1917.

Salzano, F.M. and Hurtado, A.M., Eds., *Lost Paradises and the Ethics of Research and Publication*, Oxford University Press, Oxford, 2004.

Santos, R.V., Indigenous peoples, post-colonial contexts, and genetic/genomic research in the late 20th century: a view from Amazonia (1960-2000), *Critique of Anthropology*, 22, 81–104, 2002.

Santos, R.V. and Coimbra, C.E.A., On the (un)natural history of the Tupí-Mondé Indians: bioanthropology and change in the Brazilian Amazon, in *Building a New Biocultural Synthesis: Political-Economic Perspectives on Human Biology*, Goodman, A.H. and Leatherman, T.L., Eds., University of Michigan Press, Ann Arbor, 1998, pp. 269–294.

Santos, R.V. and Coimbra, C.E.A., Cenários e tendências da saúde e da epidemiologia dos povos indígenas no Brasil, in *Epidemiologia e Saúde dos Povos Indígenas no Brasil*, Coimbra, C.E.A., Jr, Santos, R.V., and Escobar, A.L., Eds., Editora Fiocruz & Abrasco, Rio de Janeiro, 2003, pp. 13–47.

Schwarcz, L.M., *O Espetáculo das Raças: Cientistas, Instituições e Questão Racial no Brasil 1870-1930*, Companhia das Letras, São Paulo, 1993. (In English: *The Spectacle of the Races: Scientists, Institutions, and the Race Question in Brazil, 1870-1930*, Hill and Wang, New York, 1999)

Sharp, R.R. and Foster, M.W., Involving study populations in genetic research, *Journal of Law and Medical Ethics*, 28, 41–51, 2000.

Souza-Filho, C.F.M., O direito envergonhado: o direito e os índios no Brasil, in *Índios no Brasil*, Grupioni, L.D.B., Ed., Ministério da Educação e do Desporto, Brasília, 1994, pp. 153–168.

Stocking, G.W., *Race, Culture and Evolution: Essays in the History of Anthropology*, The Free Press, New York, 1968.

Todorov, T., *A Conquista da América: a Questão do Outro*, Martins Fontes, São Paulo, 1983. (In English: *The Conquest of America: The Question of the Other*, University of Oklahoma Press, Norman, 1999)

Victora, C., Oliven, R.G., Maciel, M.E., and Oro, A.P., *Antropologia e Ética: O Debate Atual no Brasil*, Editora da Universidade Federal Fluminense, Niterói, 2004.

Watkins, J., Roles, responsibilities, and relationships between anthropologists and indigenous people in the anthropological enterprise, in *Papers of the American Anthropological Association El Dorado Task Force*, American Anthropological Association, Ed., American Anthropological Association, Washington, DC, 2002, available at http://www.aaanet.org/edft/index.htm, accessed January 25, 2006.

Weijer, C., Goldsand, G., and Emanuel, E.J., Protecting communities in research: current guidelines and limits of extrapolation, *Nature Genetics*, 23, 275–280, 1999.

11

TO THE SCIENCE, TO THE LIVING, TO THE DEAD: ETHICS AND BIOARCHAEOLOGY

Bethany L. Turner, Diana S. Toebbe, and George J. Armelagos

CONTENTS

INTRODUCTION

Biological anthropologists who work with human skeletal remains frequently find themselves in a unique ethical situation. Like any other social scientists, they have an ethical responsibility to protect the interests and rights of their research subjects, but the twist for skeletal biologists, bioarchaeologists, and forensic anthropologists is that they must fulfill these responsibilities without any direct input from the subjects. The latter are deceased, some for thousands of years, and often any descendants that can be identified are separated from the deceased by many generations. On top of the difficulties this entails, skeletal biologists have to deal with an unfortunate academic legacy: the forebears of what has become the science of skeletal analysis embraced scientific racism, alienated descent populations (especially when these were poor minorities), and often treated human remains with little respect. Such behavior has not fostered trust and good will among the groups whose ancestors, once dead, were often treated as little more than objects for study — particularly Native Americans and African Americans.

Nonetheless, anthropological analyses of human remains have come a long way from their early beginnings and now make a critical contribution to the discipline. As Landau and Steele have emphasized, "human remains offer direct, tangible evidence of our history, how we have become biologically suited to the many environments in which we live, and how we behave" (1996:209). Studies of skeletal remains have been used to reconstruct population health (Steckel and Rose, 2002), sex and gender dynamics (Grauer and Stuart-Macadam, 1998), subsistence change (Cohen and Armelagos, 1984), population movement (Bentley et al., 2004; Price et al., 2000), disease transmission (Armelagos, 2004), physical activity (Kennedy, 1998), and violence (Martin and Frayer, 1997). Moreover, skeletal biologists are increasingly called upon to excavate, analyze, and contextualize remains (Grauer, 1995). For example, a team of skeletal biologists from Georgia State University were recently called upon to examine bones that had been buried in the basement of the Medical College of Georgia, and identified these as the remains of poor individuals, primarily African American, illicitly procured by the college for anatomical study (Blakely et al., 1997).

Unfortunately, any suggestion that skeletal analysis has shed its historical baggage has been undermined by a number of recent and very public events in which poor science (if not irresponsible science) has been applied to human remains, with serious ethical consequences. Indeed, however useful skeletal analysis may be, the very nature of the material studied can create contentious ethical situations for the researchers involved, whose motives might all too easily be misconstrued.

In an attempt to facilitate and foster ethical practice, the American Association of Physical Anthropologists (AAPA) recently approved an ethical code of conduct that stated that their members had "obligations to their scholarly discipline, the wider society, and the environment" (AAPA, 2003). Further, the AAPA's code:

> holds the position that generating and appropriately utilizing knowledge (i.e., publishing, teaching, developing programs, and informing policy) of the peoples of the world, past and present, is a worthy goal; that general knowledge is a dynamic process using many different and ever evolving approaches; and that for moral and practical reasons, the generation and utilization of knowledge should be achieved in an ethical manner. (AAPA, 2003)

Building on this general framework, this chapter argues that the future of skeletal analysis depends on paying close attention to several key issues. First, skeletal biologists must be fair, respectful, and compassionate not only to the remains, but also to their descendents (both definite and possible). Second, they must ensure that their conclusions do not stray beyond their data. Finally, they should be conscious of how they and their work are perceived by students, scholars from other disciplines, and the general public and how easily their methods, rationale, and interpretations can be misconstrued. Each of these issues needs to be addressed while adhering to the strictest standards of hypothesis testing, empirically based interpretations, and methodological rigor — all of which lie at the heart of the *science* of biological anthropology.

In this chapter we draw on numerous examples from the academic, legal, and popular literature to illustrate the importance of these three key issues, from high-profile controversies such as Kennewick Man, to little known historical cases such as Anthropology Days at the 1904 St. Louis World Fair. The aim of the chapter is to pay close attention to the ethical basis of research designs and methodologies that have become accepted practice in many areas of the discipline (Armelagos and Van Gerven, 2003; Baker et al., 2001; Mack and Blakey, 2004). The unifying theme of the discussion that follows is, therefore, the necessity of blurring conceptual boundaries between past and present, academia and publics, specimens and people, and fact and conjecture.

JUST MY TYPE: RACISM, PAST AND PRESENT, AND HUMAN REMAINS

American anthropology originated from, flourished in, and contributed to an intellectual climate during the early twentieth century when human evolution was believed to be the key causal mechanism underpinning social hierarchies (Baker, 1994:200). Thus, it was a time when Darwinian evolution and social Darwinism were considered one and the same — a widely accepted paradigm that provided convenient explanations (and excuses) for the racial-, gender- and class-based inequities that permeated most aspects of social life.

The 1904 St. Louis World Fair exemplified the overt racism of the time. Expositions at the fair were meant not only to entertain, but to confirm the social order to Americans who were "unsteadied by the overwhelming changes in world society" (Mathe, 1999:54). Native peoples were literally collected (Mathe, 1999) and transported to St. Louis and exhibited in what were described as their natural settings. These indigenous people were thus presented as indisputable evidence of the evolutionary superiority of the "White Races," whose technological achievements were showcased at a related exposition. Indeed, the fair advertised its exhibits as anthropological and scientifically valid, "demonstrat[ing] the progress of humanity from barbarism to the pinnacle of Anglo-Saxon civilization" (Mathe, 1999:9).

The 1904 Olympic Games also featured related "Anthropology Days." These were held in conjunction with the exposition and also sought to reinforce racial stereotypes. Individual representatives of indigenous groups from around the world were made to compete in various track and field events. W.J. McGee (in Rydell, 1987:160), one of the country's prominent anthropologists and the future first president of the American Anthropological Association (Stocking, 1960), believed this should demonstrate the evolutionary progress of humans from their "dark prime to their highest enlightenment, from savagery to civic organization, from egoism to altruism." The planned competitions, however, took an unexpected turn, and an official historian of the Olympic Games (Bennitt and Stockbridge, 1905) judged the Anthropology Days a failure because the indigenous participants failed so miserably in events such as the pole vault, the shot-put, and the weight throw* that they seriously undermined the belief that the greatest natural athletes were to be found among the uncivilized tribes of the world. But McGee promptly claimed that the competition demonstrated that civilization was not only good for the mind but also for the body.

The indignities of the Anthropology Days did not end when the games closed. Ota Benga, an Mbuti who was brought to St. Louis for the Olympics, was subsequently taken to the New York Zoo, where he was caged with a chimpanzee (Bradford and Blume, 1992). Ironically, the exhibit drew protests from the religious community, who condemned the indignity of caging a human with the chimpanzee — but only because they felt this supported evolutionary heresies.

Contemporary anthropologists are just as appalled by these practices as anyone else, but perhaps more so because the abuse was presented by its perpetrators as anthropological in nature. This does not mean that contemporary biological anthropology is free of potentially discriminatory or racist science, and there are a number of troubling aspects of contemporary skeletal research. One such aspect is the use of race in forensic

* For example, an Ainu tossed the 56 pound weight just 3′ 2″.

anthropology. Indeed, forensic anthropologists are rare among anthropologists in that they continue to recognize identifiable racial traits in skeletal remains (Gill, 1998; Sauer, 1992), despite clear evidence that race is not a biological category (Brown and Armelagos, 2001; Lewontin, 1972). The continued use of this concept in forensic anthropological research creates not only an issue of scientific integrity but also one of dishonesty in public relations. By relying on quantitative and qualitative indicators of race, forensic anthropologists are lending credence to the false but common assumption that race is a biological reality. In Norm Sauer's words, the continued use of race identification in forensic reports puts a "stamp of approval" on the traditional race concept (Sauer, 1992:110). Whether the remains are nine days or nine thousand years old, the use of race-based science in skeletal analysis harks back to a time most skeletal biologists would prefer to leave in the past.

Forensic anthropologists have a moral and professional responsibility as scientists, civil servants, and teachers to use theoretical models that reflect the limited utility of assigning individuals to races based on type specimens and typical features (Armelagos and Goodman, 1998; Goodman, 1997). Armelagos and Goodman (1998) have argued that nothing is to be gained by using a model that is known to be unsupported by data and socially divisive. The choice to identify or not to identify races is therefore not only a matter of good science, but also a matter of ethics and politics. However, it can sometimes be difficult to avoid using race, and many who are otherwise critical of the race concept find themselves supporting its use in forensic contexts. Some no doubt subscribe to the view that they are simply providing an opinion that is analogous to an eyewitness report of a suspect's observed race. This approach permits forensic anthropologists to provide more information about the skeletal material they are analyzing — something that is no doubt gratifying in and of itself. It also lets anthropologists off the hook of racism, because calling them racists for using racial classifications would be the same as calling the eyewitness a racist. In situations where police officers say they require racial information to locate or identify a missing person, refusing to offer a racial assessment might seem ludicrous.

However, the problem with this argument is that the anthropologist and the eyewitness are not equivalent. Anthropologists are counted as experts, and their words and deeds are endowed with all the symbolic authority that science can bestow. Thus, when forensic anthropologists make racial identifications of skeletal remains, the information they provide is presumed to be more accurate than those provided by lay eyewitnesses, even though the anthropologists might have had access to substantially less data than the eyewitnesses. Moreover, when anthropologists assign race, they create the impression that race is something that *can* be scientifically identified. While one could argue that the identification of race has allowed police forces around the world to make a higher percentage of correct identifications, it is impossible to know how many individuals have gone unidentified or have been misidentified because the racial labels attached to them by forensic anthropologists were different than those recognized by the authorities or society at large — i.e. their bureaucratic race. Moreover, while race has traditionally been one of the four criteria that make up the standard biological profile of forensic reports (alongside age, sex, and stature), it is not required information for identifying an individual. For this reason the jury is out on whether the use of race in forensic science is more of a help than a hindrance.

POSTHUMOUS RIGHTS: THE STUDY AND REPATRIATION OF NATIVE AMERICAN REMAINS

During the nineteenth and twentieth centuries, the burial grounds of Native Americans were routinely disturbed in order to collect both cultural artifacts and human remains for sale and analysis (Beider, 1996; Peterson, 1990/1991). A common form of analysis was the identification and measurement of cranial features presumed to differ between Europeans and other races, such as those carried out by Samuel Morton (Beider, 1996) and Paul Broca (Gould, 1996). These data were used to quantify race-based levels of intelligence, morality, and position in a unilinear evolutionary hierarchy — a paradigm that persisted well into the twentieth century and the dawn of Sherwood Washburn's New Physical Anthropology (Armelagos and Van Gerven, 2003).

These scientific agendas, which are not disconnected from political and socioeconomic agendas, motivated scientists (and the grave robbers they occasionally employed) to scour the American landscape for the crania of different races. Crania were collected from graves of varying ages and contexts, sometimes severed from recently buried corpses, and Native American graves were those most frequently targeted (Thomas, 2000). The data upon which such sweeping and inherently biased generalizations were based were themselves biased — based on methods corrupted by the *a priori* assumptions of the scholars carrying out the research (Gould, 1996), in which White Europeans stood at the zenith of racial hierarchies, while Native Americans found themselves at the bottom.

The practice of grave intrusion and the disproportionate collection of Native American remains was neither condemned by anthropology nor relegated to a racist fringe. For example, Franz Boas, the celebrated founder of North American anthropology, presided over the funeral of a Native Eskimo study subject and lent support to the man's orphaned son, having secretly arranged beforehand to remove the body from its casket for research (Thomas, 2000). Boas' motivation likely stemmed from the meticulously descriptive nature of his research and his interest in measuring variation among human populations. Indeed, Boas stood in firm opposition to the notion of immutable racial characterizations (Boas, 1940) and was motivated more by an interest in preserving the history and culture of what he, like many other scholars of the nineteenth and twentieth centuries, saw as a dying people (Thomas, 2000). Native Americans were romanticized as polarized against the modern world, their eventual extinction viewed as an unfortunate but inevitable result of their inability to assimilate into Anglo-American culture (Beider, 1996; Hinsley, 1996). At the time a preoccupation with recording the characteristics of these dying peoples would certainly have justified Boas' actions, but in hindsight they appear outrageous and reprehensible.

Such behavior is not relegated to the past. As recently as the 1970s, the discovery of cemeteries holding White, African American, and Native American skeletal remains often resulted in the relocation and reburial of the former two and the transportation of the latter to museums or skeletal biology laboratories (Bowman, 1989; Dongoske, 1996). Owing to shortages in funding and researchers, there was often no guarantee that the remains would even be analyzed (Rose et al., 1996), and there was rarely any meaningful consultation with descendants. Attempts by Native American tribes to regain control over ancestral remains were complicated by grave protection policies that were, at best, weakly enforced and often placed the burden on the tribes to prove that repatriation was worth the loss of scientific information (Bowman, 1989). Native American burial grounds, especially those of great antiquity, were often poorly marked and lacked features, such as headstones, that would have helped them qualify as legal cemeteries. Moreover, tribes that had been forcibly relocated by other tribes or by European colonizers often lost track of cemetery locations or lost their claims to ancestral lands (Peterson, 1990/1991). Finally, under the Archaeological Resources Protection Act of 1979, Native Americans were allowed greater participation in granting permits for excavations on federal lands. However, the deciding vote as well as ownership of any excavated materials and remains rested with the federal government (Bowman, 1989).*

* To provide a context to these issues, it is important to consider that Native Americans did not become citizens until 1924 and waited even longer for the right to vote and only in 1978 were officially given the right to perform indigenous religious ceremonies, access sacred sites, and use sacred objects under the American Indian Religious Freedom Act (Peterson, 1990/1991). Even this concession was circumvented on numerous occasions, as legal loopholes were employed to limit the religious freedoms of tribes when such freedoms applied to land rights (Bowman, 1989).

Nonetheless, many anthropologists have been drawn to skeletal analysis because they are keen to explore the injustices suffered by such populations in the past. American biological anthropology is often strongly materialist and tends to be focused on political–economic processes in its studies of health and well-being in the past and present — explicitly emphasizing the effects of power inequities and differential resource access on health outcomes in contemporary, historic, and ancient populations (Armelagos et al., 2004; Goodman and Leatherman, 1998; Goodman et al., 1995). This is especially true for researchers whose work focuses on Native American populations. Here, there is a great deal of interest in reconstructing the health, activities, and demography of indigenous populations, including the adversities they met during periods of cultural transition and following European contact (Ericksen, 1980; Goodman and Armelagos, 1985; Gordon et al., 2003; Lallo et al., 1980; Larsen et al., 2001; Larsen and Milner, 1994; Martin and Goodman, 2002; Schoeninger et al., 1987; Steyn, 2003; Ubelaker et al., 1995). Indeed, given the traditional marginalization of many indigenous populations in mainstream studies of world history (Wolf, 1982), there is added value in research that adds to what little is known of their past.

To this end, many bioarchaeologists have maintained positive collaborative relationships with indigenous descent communities and have striven to make their research both ethically balanced and socially relevant (Anyon et al., 1997; Baker et al., 2001; Martin and Frayer, 1997). In 1990, with the passage of the U.S. Native American Graves Protection and Repatriation Act (Public Law 101-601; commonly referred to as NAGPRA), the involvement of U.S. indigenous descent communities in skeletal analysis moved from ethical imperative to legal mandate. Any institution receiving federal funding, including museums and universities (public and private), had to compile inventories of all of the Native American cultural materials and human remains housed in their facilities and had to make these publicly available. Skeletal remains and associated items from graves excavated on federal lands with an established affiliation to one or more living Native American descent community can now be repatriated to those communities upon request and, if so desired by tribal members, re-interred (Rose et al., 1996). In addition, any development projects or excavations that accidentally or intentionally discover human remains determined to be Native American have to cease work pending notification of NAGPRA officials and the decisions of local tribes (Lannan, 1998; Rose et al., 1996). Scientific analyses may no longer be done on remains excavated from tribal lands without the prior consent of the tribe, although if these remains are excavated on federal land, analyses may be undertaken so long as the tribes concerned are consulted (Rose et al., 1996).

Some of the concerns expressed by some bioarchaeologists over the unnecessary repatriation of skeletal remains seem reasonable, given the advances resulting from the reexamination of skeletal populations using new methods and theoretical frameworks (Buikstra and Gordon, 1981). Others have found it more productive to view NAGPRA as human rights legislation rather than as a set of restrictive regulations (Tsosie, 1997). As such, NAGRA is not merely an opportunity to right historical wrongs but also to shape the future of research (Fine-Dare, 2002). NAGPRA is, in effect, federal recognition of the rights of Native Americans to decide what becomes of the remains and grave accompaniments of their ancestors — a right seldom considered for White and African American remains, but seldom necessary either.

As different as the interests of Native Americans and skeletal biologists are, the necessary reporting and possible repatriation of remains stipulated by NAGPRA has nonetheless resulted in unexpected benefits for skeletal research. Indeed, Rose and colleagues (1996) argue that the legislation has revitalized human osteology in a number of ways, a notion echoed by others in the field (Dongoske, 1996). These include the standardization of data collection methods and the coding of large-scale osteological databases (Buikstra and Ubelaker, 1994) and the comprehensive analyses of long-unstudied skeletal collections. Both have rejuvenated skeletal biology, making data sets more directly comparable and filling many gaps in knowledge through increased research activity. Moreover, discoveries of new Native American remains have not inevitably led to reburial without analysis. The Hopi, Zuni, and Maricopa, for example, have continually expressed interest in, and maintained involvement with, the analysis of affiliated skeletal remains (Dongoske, 1996; Ferguson, 1996; Ravesloot, 1997). Indeed, a substantial proportion of Native Americans have expressed the belief that archaeology and bioarchaeology are beneficial to the preservation of Native American culture (Klessert and Holt, 1975, cited in Ferguson, 1996).

However, the unexpected benefits of NAGPRA for bioarchaeology are not the bottom line for descent populations, whose primary concern is for the proper treatment and appropriate care of their ancestors. Although their concern is strong, it is not indiscriminate. In several case studies descendants have wanted to firmly establish the cultural affiliation of remains before claiming them, so as not to inter ancestral enemies on tribal land or perform sacred rituals on the ancestors of others (Cantwell, 2000). Since museums and universities are equally reluctant to repatriate valued collections when tribal affiliations are uncertain, there is a mutual interest in establishing the cultural identity of any material using what NAGPRA defines as a "preponderance of evidence," without limiting what form that evidence should take (Johnson, 2002).

Given the intentional vagueness of the NAGPRA criteria, there has been an inherent power dynamic in the assessment of what evidence is most accurate or authentic for establishing cultural affiliation (Johnson, 2002). This is especially true in the assessment of poorly preserved remains, remains with incomplete accession information, and ambiguous burial contexts. A disproportionate reliance on biological criteria (such as patterns of dental morphology, discrete traits, craniometric measurements, and overall stature) can foster tension between researchers and communities that give equal or greater weight to oral histories, religious beliefs, and ceremonial ritual as explanatory mechanisms of cultural association (Anyon et al., 1997; Cantwell, 2000; Ferguson, 1996; Zimmerman, 1997). Of course one would be foolish to assume that the motivation behind a reliance on one or another type of evidence is purely objective for either researchers or concerned tribes.

An infamous example is the controversy surrounding the cultural affiliation of a 9300-year-old, well-preserved man discovered by the Columbia River in Washington — commonly referred to as Kennewick Man. The antiquity, ambiguous context, and exceptional preservation of Kennewick Man made the discovery, and the ensuing debate over his cultural affiliation, a high-profile affair. A coalition of Native American tribes successfully put forward a claim for Kennewick Man and requested immediate repatriation from the Army Corps of Engineers (who had jurisdiction over the land and were in possession of the remains). This did not sit well with several skeletal anthropologists, who in turn filed a suit against the Corps, citing repatriation without first establishing cultural affiliation as a NAGPRA violation and demanding permission to study the remains (*Bonnishsen et al. vs. the United States*; Lannan, 1998). Further fuelling the debate, several researchers independently arrived at the conclusion that Kennewick Man exhibited "Caucasoid" traits, often interpreted as indicative of European ancestry, in contrast to the "Mongoloid" features characteristic of Asian descent (Lannan, 1998).

Despite the absence of any evidence of European arrival to the Western hemisphere prior to the Common Era, some have suggested, based on these data, that Kennewick Man is actually of European descent. Such a contention has not only sought to invalidate the claims of local tribes over the remains, but created a firestorm of competing claims from religious groups such as the Asatru Folk Assembly of California and several white supremacist groups,* impacting public perceptions of tribal land claims and of Native American identity (Lannan, 1998). This misconception

* For example, http://www.pbs.org/wotp/kennewick_man/ and http://www.newnation.org/NNN-kennewick-man.html, accessed January 25, 2006.

persists despite conclusive evidence that the Kennewick Man skeleton is "clearly not a Caucasoid," but rather shares morphological similarities with Polynesian and East Asian groups such as the Ainu, which were described as "Caucasoid" by anthropologists in the nineteenth century (Powell and Rose, 1999). Those morphological features of Kennewick Man that are shared with populations of European ancestry are also shared with populations of Asian ancestry — significantly weakening any definitive claims of his population affiliation. Moreover, his greater similarity to East Asian populations versus more recent Native American populations is not surprising, given the extreme antiquity of the Kennewick skeleton relative to most Native American skeletal populations (Powell and Rose, 1999). The recent ruling, in *Bonnichsen et al. vs. the United States*, that Kennewick Man cannot be affiliated to any known Native American tribe, and thus does not fall under NAGPRA's guidelines, will inevitably have substantial implications for the way in which remains of extreme antiquity are dealt with in the future (Holden, 2004).

The controversy surrounding Kennewick Man, while not representative of most repatriation cases, illustrates a key component of the emerging relationship between biological anthropologists and Native American communities — that of identity. It is interesting how, for all of its complications and shortcomings, NAGPRA has unilaterally and unexpectedly succeeded in forcing researchers and descent communities to modify the collective identities that they present to each other and the public at large. Biological anthropologists must confront the well-founded mistrust expressed by indigenous and other minority groups in order to defend the credibility, integrity, and value of their research (White Deer, 1997; Zimmerman, 1997). Indeed, many biological anthropologists feel that the just and proper course of action has often been to re-inter Native American remains. Many have themselves participated in the ceremonies, knowing full well that they are losing any future opportunities to study the remains (Baker et al., 2001).

Native American communities, in turn, must grapple with conflicting aspects of their own identities. While there is a deeply held respect for the traditions and sanctity of ancestral bodies and ways of life, these traditions may be one of several overlapping ideologies of descendants who are Christians, scientists, and even archaeologists (Lippert, 1997). In the case of ambiguously affiliated or unaffiliated remains, decisions must be made as to whether sacred ceremonies should be performed on individuals who may not be ancestors. They must also wrestle with maintaining intra- and intertribal diversity, while presenting a unified set of wishes and expectations to NAGPRA officials regarding ancestral remains (Johnson, 2002), if not a definition of what constitutes a "real" Native American (Cantwell, 2000).

Such dilemmas have often been positively resolved through open dialogue, and this ongoing and dynamic shaping of individual and group identities seems to be a sign of progress as biological anthropologists and native communities move ever closer to greater understanding. The controversy over Kennewick Man has shown, however, that situations in which the interests and stakes are high for both groups can quickly undermine positive relations. This is especially the case when one or both sides rely on ethically questionable, ambiguous, or fuzzy sources of evidence to support their claims, such as the use of temporally and environmentally sensitive cranial criteria to establish ancestry (Goodman, in Holden, 2002). The use of these criteria harks back to a long history of unethical, socially and politically motivated scientific inquiry in American anthropology that was not confined to the iniquitous treatment of Native Americans. While the Kennewick case is certainly one of the most widely known, it is merely one of a number of examples that demonstrate the pernicious resilience of such ideologies.

ISSUES IN THE CLASSROOM AND AMONG THE GENERAL PUBLIC

At the root of much that has been discussed in this chapter is the notion that many ethically contentious situations are made so by a failure to uphold fair and equal standards of treatment across all skeletal populations and concerned descent groups. Many skeletal biologists have worked to remedy these practices through open, fair-minded, and respectful relationships with descent groups and by adhering to strict standards of ethical behavior. However, to eradicate unethical practice in skeletal research, these standards should be coupled with a healthy dose of caution when interpreting and presenting data. Much information can be gained from skeletal analyses, but audiences that are less familiar with the sorts of limitations that are well understood by more seasoned researchers can easily misinterpret results, especially when these are presented in an ambiguous or cavalier fashion.

Just as skeletal biologists have an ethical responsibility to their colleagues and to descent communities, they have a responsibility to those with little to do with skeletal remains — to the general public, among them undergraduate students, who often have no previous experience of dealing with preserved skeletal material. Skeletal biologists and bioarchaeologists often have excellent opportunities to present their research to popular audiences and in smaller, hands-on classroom settings, but skeletal biology is no different from any other academic science in that scholars have an obligation to be honest and truthful with coworkers and students inside and outside the classroom.

The teaching of bioarchaeology and skeletal biology brings certain unconventional ethical challenges to classroom and university settings. College students are likely to enroll in skeletal biology classes to learn something they do not already know about a topic relatively peripheral to mainstream undergraduate education, and the professor must wrestle with the need to make the course material interesting and engaging without encouraging the use of outdated methods and debunked theory. For example, it is argued here that it is ethically irresponsible to teach students the methods of morphological typology if such a lesson does not emphasize the flaws and questionable ethics of such research. When students learn to "race" skulls, they are learning to observe features and then pigeonhole individuals according to a list of typical traits. They learn nothing about the evolutionary processes that led to variation in humans or about the distribution of such traits at a population level, and given the poor current job market for forensic anthropologists, it is increasingly unlikely that these students will go on to apply their knowledge to identifying skeletons. Instead, they will take what they have learned and apply it elsewhere, either in anthropology, some other part of academia, the business world, or medicine, perpetuating a false paradigm of bounded, natural races with inherent differences in ability (Entine, 2000; Goodman, 1997).

Another check on the desire to make classroom subject matter interesting and relatable to students is the risk of stretching the boundaries of our knowledge past the point of scientific reasoning. Researchers often push the scholarly envelope to expand upon what is known, going beyond what are traditionally considered standard theories, methods, and applications. It is this willingness to wander into uncertain territory that makes science relevant, exciting, and vibrant. However, when asked to apply pertinent knowledge to practical situations in legal or emotionally charged contexts, skeletal biologists and forensic anthropologists are ethically bound to offer advice framed by the standards and practices established in their field. In their widely used textbook, *Human Osteology*, White and Folkens (2000) provide an excellent example of how forensic investigators with the best of intentions can go beyond the boundaries of their knowledge. A decade after the crash of a C-130 gunship shot down in Laos in 1972, the excavation of the crash site uncovered some 50 thousand bone fragments, each measuring 1 to 13 cm in length. U.S. Army forensic experts identified 13 of the victims from these fragments, assigning sex, age, and race to each. Yet in skeletal biology it is widely accepted that each of these characteristics are assessed using multiple diagnostic criteria from identifiable skeletal elements (Buikstra and Ubelaker, 1994; Lovejoy et al., 1985), and even then, exact assessments can be difficult. In the case of race, the validity of *any* assessment is questionable (Brown and Armelagos, 2001). Given the state of the human remains at the crash site, the identification of these complex characteristics was clearly questionable. After complaints from the victims' relatives, independent professional skeletal biologists repeated the analyses and concluded that the identifications performed by the army team were inappropriate, given the nature of the osteological data and the emotional significance to the next of kin. The use of such an example in the classroom is an excellent way to illustrate the importance of balancing scientific enthusiasm with a healthy dose of caution and pragmatism.

High-profile, high-stakes examples such as these can capture the attention of students and effectively raise an awareness of how prudent it is to be rigorous and cautious — and how deplorable it can be to stretch one's data and interpretations beyond what is reasonable. Imparting a sense of ethical responsibility in the skeletal biology classroom can also be accomplished quite effectively with something as simple as a closer examination of seemingly neutral classroom tools. For instance, many of the skeletons in osteology labs were purchased decades ago from anatomical supply houses and were often impoverished individuals (Blakely et al., 1997). This is an unfortunate, indeed ugly, fact that many anthropologists would surely prefer to forget. However, to do so would not only be remiss, but would result in a missed opportunity to drive home abstract ethical concepts to students, using a concrete example that is right in front of our noses.

Moreover, the social and political economic issues pertinent to contextualizing skeletal studies in the classroom are not confined to the pages of history books or to bones acquired decades ago. Much of the racism, discrimination, socioeconomic marginalization, poor health, and violence written in the bones of the skeletons under study are still experienced, with varying degrees, by current Native American (Cantwell et al., 1998; Chester et al., 1999; De Coteau et al., 2003; Dillinger et al., 1999; Flores et al., 1999; Grossman et al., 1991; Harris and Harper, 2001; Lester, 1995; May, 1992; McGinnis and Davis, 2001; Santosham et al., 1995) and African American populations (de Groot et al., 2003; Evans, 2004; Geronimus, 2003; Gwyn et al., 2004; Lichtenstein, 2003; Mellor and Milyo, 2004; Minior et al., 2003; Sankar et al., 2004). The ethical responsibility inherent in the study of human remains will perhaps be clearer to undergraduate and graduate students if both historical and contemporary issues are interwoven into osteological course materials. Indeed, if students do not learn the historical and contemporary issues that are inextricably linked to the skeletal populations being studied, then no matter how extensive their anatomical, histological, and diagnostic knowledge, their training will be incomplete.

Moving beyond the classroom, the scope and character of the interaction between academic scientists and the general public is woefully under-theorized yet extremely important. For bioarchaeologists, the way the public perceives our role and contribution is key. Yet many bioarchaeologists and skeletal biologists may never be *formally* taught how to avoid the potential ethical pitfalls that await them when they venture outside academia. With the exception of experienced forensic anthropologists, seasoned NAGPRA consultants, and contract archaeologists, anthropologists may be unprepared for the high level of emotion attached to skeletal remains by those who do not handle and study them on a daily basis (Baker et al., 2001; Mack and Blakey, 2004). Without a thorough understanding of the tremendous symbolic meanings that skeletal remains hold for many people, bioarchaeologists might still be viewed by descent communities as crass, at best, and criminal, at worst, even if they adhere to the rules of their discipline. It is as if by virtue of their training and continued experience, an entire group of professionals has become desensitized to a set of emotions universal to the rest of the world. Just as doctors have been rightly criticized for a perceived lack of empathy when dealing with the afflictions of their patients (Konner, 1987), bioarchaeologists are open to similar criticisms because they appear so comfortable dealing with human remains.

This is especially the case for the remains of the recently deceased, those with strong tribal or group affiliations, and prominent historical figures and is not confined to the remains of the historically disenfranchised. A personal anecdote might provide a helpful illustration. One of the authors (D.S.T.) was a member of a team asked to analyze a skeleton that had been discovered in the grounds of Roaring Camp Railroad historical park in northern California in 1995. When the remains were examined along with the artifacts found associated with them (including coins, a gun, and a whiskey flask), it was clear that this was an individual from the late 1800s, and not a recent murder or a prehistoric Native American burial. The result of the examination was a poster presented at the national meeting of the American Association of Physical Anthropologists that was received as a interesting find but of little theoretical value (Smay et al., 1998). Having wrapped up the excavation and presented their results, the members of the team assumed that their job was finished, but once the presentation was completed, the Roaring Camp Railroad staff retained an unexpected attachment to the remains. The park is itself a historical reconstruction of a frontier post of the 1880s, and many of the reenactment staff were eager to give "Willy" (as they had dubbed the skeleton) a proper historic burial. The archaeologists were glad to help but were completely unprepared for what was to follow. Hundreds of people, many in period costume, turned out for the burial ceremony. Most of the individuals were not formally affiliated with the park but simply had an avid interest in regional history in general and Willy in particular. The event consisted of a parade, bagpipe serenade, 21-gun salute, passionate eulogy, and formal funeral. Newspaper and television news reporters covered the event, and the remains — which had been seen, primarily, as a source of data, a modest scientific curiosity, and the basis for an academic presentation — took on an entirely different meaning for the assembled mourners.

A similar clash of the academic and the public worldview led to the tensions that arose at the African Burial Ground in New York City (Blakey, 1998; Epperson, 1996; LaRoche and Blakey, 1997; Mack and Blakey, 2004). In an event that was a wake-up call for many bioarchaeologists, the African-American population of New York and elsewhere in the country made it clear that they would not tolerate what they felt was the academic sanitization of history. Ironically, the archaeologists involved were interested in investigating the health of African American slaves, but in the eyes of the African American community, the wholesale excavation of their ancestors was paternalistic, inexcusable, and racist. Interactions between researchers and African American communities in Dallas and Newark, while less contentious and lower-profile, also emphasize the gulf between those who see human skeletal remains as important sources of information and those for whom such remains hold strong emotional salience (Roberts and McCarthy, 1995).

It would seem necessary, then, to allow public opinion to trump the scientific value of human remains — particularly in the face of potentially harmful impacts on relatives, descendants, and public perception (Baker et al., 2001; Roberts and McCarthy, 1995). This is, of course, easier said than done when the average researcher spends years devoted to field and laboratory training and scholarly investigation, but as anthropologists first and foremost, skeletal biologists and bioarchaeologists are better equipped than most to be sensitive to the ways in which multiple value systems may diverge, intersect, and conflict. They are well equipped to mediate between the general public and marginalized groups and to dispel common misconceptions regarding the methods and subjects of anthropological research (Baker et al., 2001; Bowman, 1989; Ferguson, 1996; Peterson, 1990/1991). Yet all biological anthropologists are scientists, loyal to their own set of perspectives, codes of conduct, and modes of action. As specialists with uniquely holistic training, skeletal biologists and bioarchaeologists will continue to negotiate the science and the meaning of what they do, how they do it, how they teach it, and how they successfully interface with other interested parties as new methods develop, new discoveries are made, and older ones are revisited.

RECOMMENDATIONS

Human beings have the capacity to act ethically and yet are also fallible — prone to taking the path of least resistance. This has translated into a history of lax ethics in the treatment of human skeletal remains and to a good deal of uncertainty as to the future of skeletal research. As skeletal scientists continue to develop their discipline, most recognize that they need to do more than simply identify areas of inconsistent ethical practice, acknowledge historical baggage, and accept differences in perspective. They need to develop new ways of working to forge better relations within the broader anthropological community and with concerned segments of the general public. The following recommendations summarize some of the key options available to skeletal scientists:

Seek Continual Dialogue and Reflective Critique

Bioarchaeologists, skeletal biologists, and scholars from related disciplines should continue to engage in self-reflection, ethically minded peer critique, and open dialogue to make all aspects of skeletal excavation and analysis scientifically rigorous, theoretically relevant, and ethically conscious. Collaborative efforts to codify a set of guidelines for reference in ambiguous or complicated contexts would help all parties minimize contentious situations and establish valuable dialogue between parties with different worldviews (Johnson, 2002).

Encourage Involvement and Training of Members of Historically Marginalized Groups

The involvement of concerned groups should not begin and end with consultation. Colleges, universities, museums, and archaeological firms should seek to involve members of descent communities (and particularly the historically disenfranchised) in all stages of research — from excavation and cataloguing to storage, analysis, and publication. Such involvement would provide not only opportunities to present the fundamental concepts and methods of bioarchaeology and skeletal biology to descent communities more directly but also opportunities to incorporate diverse perspectives in shaping the treatment and analysis of human remains.

Focus Increasingly on Cross-Cultural Treatment of Indigenous Remains

Beyond the ethical concerns and historical underpinnings of American bioarchaeology, increased attention should be paid to the ongoing ethical concerns relating to the excavation, curation, and analysis of indigenous skeletal remains in other countries. The effects of different political economic histories in countries such as those in Central and South America, Australia,* and

* See: Australian Aboriginal and Torres Strait Islander Heritage Protection Act 1984 http://www.austlii.edu.au/au/legis/cth/consol_act/aatsihpa1984549/ and Protecting the Past http://www.cr.nps.gov/seac/ protecting/index.htm, accessed January 25, 2006.

elsewhere could further elucidate analogous effects in other locations and inform wider policies on the classification and treatment of human remains.

Acknowledging and accepting the systems of meaning among Native American nations, African American communities, and the general public at large is also imperative. Equally important is adhering to theoretical and methodological rigor. As perhaps the most holistic of the sciences, anthropology seems uniquely suited to integrate scientific and alternative modes of social knowledge into a workable framework.

REFERENCES

AAPA, *Code of Ethics of the American Association of Physical Anthropologists*, 2003, available at http://www.physanth.org/positions/ethics.htm, accessed January 2006.

Anyon, R., Ferguson, T.J., Jackson, L. et al., Native American oral tradition and archaeology: issues of structure, relevance, and respect, in *Native Americans and Archaeologists: Stepping Stones to Common Ground*, Swidler, N., Dongoske, K.E., Anyon, R., et al., Eds., AltaMira Press, Walnut Creek, CA, 1997, pp. 77–87.

Armelagos, G.J., Emerging disease in the third epidemiological transition, in *The Changing Face of Disease: Implications for Society*, Mascie-Taylor, N., Peters, J., and McGarvey, S.T., Eds., CRC Press, Boca Raton, FL, 2004, pp. 7–23.

Armelagos, G.J., Brown, P.J., and Turner, B.L., Evolutionary, historical and political economic perspectives on health and disease, *Social Science and Medicine*, 61, 755–765, 2005.

Armelagos, G.J. and Goodman, A.H., Race, racism and anthropology, in *Building a New Biocultural Synthesis: Political-Economic Perspectives on Human Biology*, Goodman, A.H. and Leatherman, T.L., Eds., University of Michigan Press, Ann Arbor, 1998, pp. 359–377.

Armelagos, G.J. and Van Gerven, D.P., A century of skeletal biology and paleopathology: contrasts, contradictions, and conflicts, *American Anthropologist*, 105, 53–64, 2003.

Baker, B.J., Wilkinson, R.G., Varney T.L. et al., Repatriation and the study of human remains, in *The Future of the Past: Archaeologists, Native Americans, and Repatriation*, Bray, T.L., Ed., Garland, New York, 2001, pp. 69–89.

Baker, L.D., The location of Franz Boas within the African-American struggle, *Critique of Anthropology*, 14, 199–217, 1994.

Beider, R.E., The representations of Indian bodies in nineteenth-century American anthropology, *American Indian Quarterly*, 20, 165–180, 1996.

Bennitt, M. and Stockbridge, F.P., *History of the Louisiana Purchase Exposition: St. Louis World's Fair of 1904*, Universal Exposition Publishing Company, St. Louis, MO 1905.

Bentley, R.A., Price, T.D., and Stephan, E., Determining the 'local' Sr-87/Sr-86 range for archaeological skeletons: a case study from Neolithic Europe, *Journal of Archaeological Science*, 31, 365–375, 2004.

Blakely, R.L., Harrington, J.M., and Barnes, M.R., *Bones in the Basement: Postmortem Racism in Nineteenth-Century Medical Training*, Smithsonian Institution Press, Washington, DC, 1997.

Blakey, M.L., The New York African Burial Ground project: an examination of enslaved lives, a construction of ancestral ties, *Transforming Anthropology*, 7, 53–58, 1998.

Boas, Franz, *Race, Language and Culture*, Free Press, New York, 1940.

Bowman, M.B., The reburial of Native American skeletal remains: approaches to the resolution of a conflict, *Harvard Environmental Law Review*, 13, 147–208, 1989.
Bradford, P.V. and Blume, H., *Ota Benga: the pygmy in the zoo*, St. Martin's Press, New York, 1992.
Brown, R.A. and Armelagos, G.J., Apportionment of racial diversity: a review, *Evolutionary Anthropology*, 10, 15–20, 2001.
Buikstra, J.E. and Gordon, C.C., The study and restudy of human skeletal series: the importance of long-term curation, *Annals of the New York Academy of Sciences*, 376, 449–465, 1981.
Buikstra, J.E. and Ubelaker, D.H., Standards for data collection from human skeletal remains, in *Proceedings of a Seminar at the Field Museum of Natural History*, Archaeological Survey Research Series No. 44, Fayetteville, AR, 1994.
Cantwell, A.-M., Who knows the power of his bones? Reburial redux, *Annals of the New York Academy of Sciences*, 925, 79–119, 2000.
Cantwell, M.F., et al., Tuberculosis and race/ethnicity in the United States: impact of socioeconomic status, *American Journal of Respiratory and Critical Care Medicine*, 154, 1016–1020, 1998.
Chester, B.P., Mahalish, P., and Davis, J., Mental health needs assessment of off-reservation American Indian people in northern Arizona, *American Indian and Alaska Native Mental Health Research*, 8, 25–40, 1999.
Cohen, M.N. and Armelagos, G.J., Eds., *Paleopathology at the Origins of Agriculture*, Academic Press, Orlando, FL, 1984.
De Coteau, T.J., Hope, A., and Anderson, J., Anxiety, stress, and health in northern plains Native Americans, *Behavior Therapy*, 34(3), 365–380, 2003.
de Groot, M., et al., Depression and poverty among African American women at risk for type 2 diabetes, *Annals of Behavioral Medicine*, 25, 172–181, 2003.
Dillinger, T.L., Jett, S.C., Macri, M.J., et al., Feast or famine? Supplemental food programs and their impacts on two American Indian communities in California, *International Journal of Food Sciences and Nutrition*, 50, 173–187, 1999.
Dongoske, K.E., The Native American Graves Protections and Repatriations Act: a new beginning, not the end, for osteological analysis: a Hopi perspective, *American Indian Quarterly*, 20, 287–297, 1996.
Entine, J., *Taboo: Why Black Athletes Dominate Sports and Why We're Afraid to Talk about It*, Public Affairs, New York, 2000.
Epperson, Y.W., The politics of 'race' and cultural identity at the African Burial Ground excavations, New York City, *World Archaeological Bulletin*, 7, 108–117, 1996.
Ericksen, M.F., Patterns of microscopic bone remodelling in three Aboriginal American populations, in *Early Native Americans: Prehistoric Demography, Economy and Technology*, Brownman, D.L., Ed., Mouton Publishers, The Hague, 1980.
Evans, G.W., The environment of childhood poverty, *American Psychologist*, 59, 77–92, 2004.
Ferguson, T.J., Native Americans and the practice of archaeology, *Annual Review of Anthropology*, 25, 63–79, 1996.
Fine-Dare, K.S., *Grave Injustice: The American Indian Repatriation Movement and NAGPRA*, University of Nebraska Press, Lincoln, 2002.
Flores, G., Bauchner H., Feinstein, A.R. et al., The impact of ethnicity, family income, and parental education on children's health and use of health services, *American Journal of Public Health*, 80, 1066–1071, 1999.
Geronimus, A.T., Damned if you do: culture, identity, privilege, and teenage childbearing in the United States, *Social Science and Medicine*, 57, 881–893, 2003.

Gill, G.W., Craniofacial criteria in the skeletal attribution of race, in *Advances in Identification of Human Remains*, 2nd ed., Reichs, K.J., Ed., Charles C. Thomas, Springfield, IL, 1998, pp. 293–317.

Goodman, A.H., Bred in the bone? *Sciences,* March/April, 20–25, 1997.

Goodman, A.H. and Armelagos, G.J., Disease and death at Dickson Mounds, *Natural History*, 94, 12–18, 1985.

Goodman, A.H. and Leatherman, T.L., Eds., *Building a New Biocultural Synthesis: Political-Economic Perspectives on Human Biology*, University of Michigan Press, Ann Arbor, 1998.

Goodman, A.H., Martin, D.J., and Armelagos, G.J., The biological consequences of inequality in prehistory, *Rivisita di Anthropologia*, 73, 123–131, 1995.

Gordon, K.D., Schoeninger, J., and Sears, K.E., Diet in pre-contact central California explored through dental microwear and stable isotope analyses, *American Journal of Physical Anthropology*, 122, 102–103, 2003.

Gould, S.J., *The Mismeasure of Man*, Norton, New York, 1996.

Grauer, A.L., Ed., *Bodies of Evidence: Reconstructing History through Skeletal Analysis*, Wiley-Liss, New York, 1995.

Grauer, A.L. and Stuart-Macadam, P., Eds., *Sex and Gender in Paleopathological Perspective*, Cambridge University Press, Cambridge, 1998.

Grossman, G.C., Milligan, B.C., and Deyo, R.A., Risk factors for suicide attempts among Navajo adolescents, *American Journal of Public Health*, 81, 870–874, 1991.

Gwyn, K., Bondy, M.L., Cohen, D.S., et al., Racial differences in diagnosis, treatment, and clinical delays in a population-based study of patients with newly diagnosed breast carcinoma, *Cancer*, 100, 1595–1604, 2004.

Harris, S. and Harper, B.L., Lifestyles, diets, and Native American exposure factors related to possible lead exposures and toxicity, *Environmental Research*, 86, 140–148, 2001.

Hinsley, C.M., Digging for identity: reflections on the cultural background of collecting, *American Indian Quarterly*, 20, 180–197, 1996.

Holden, C., Going head-to-head over Boas's data, *Science*, 298, 942–945, 2002.

Holden, C., Kennewick Man: court battle ends, bones still off-limits, *Science*, 305, 591, 2004.

Johnson, G., Tradition, authority and the Native American Graves Protection and Repatriation Act, *Religion*, 32, 355–381, 2002.

Kennedy, K.A.R., Markers of occupational stress: conspectus and prognosis of research, *International Journal of Osteoarchaeology*, 8, 305–310, 1998.

Konner, M., *Becoming a Doctor: A Journey of Initiation in Medical School*, Viking, New York, 1987.

Lallo, J., Rose, J.C., and Armelagos, G.J., An ecological interpretation of variation in mortality within three prehistoric American Indian populations from Dickson Mounds, in *Native Americans: Prehistoric Demography, Economy and Technology*, Brownman, D.L., Ed., Mouton, The Hague, 1980, pp. 203–238.

Landau, P.M. and Steele, D.G., Why anthropologists study human remains, *American Indian Quarterly*, 20, 209–229, 1996.

Lannan, R.W., Anthropology and restless spirits: The Native American Graves Protection and Repatriation Act, and the unresolved issues of prehistoric human remains, *Harvard Environmental Law Review*, 22, 369–384, 1998.

Larsen, C.S., Griffin, M.C., Hutchinson, D.L., et al., Frontiers of contact: bioarchaeology of Spanish Florida, *Journal of World Prehistory*, 15, 69–123, 2001.

Larsen, C.S. and Milner, G.R., Eds., *In the Wake of Contact: Biological Responses to Conquest*, Wiley-Liss, New York, 1994.

LaRoche, C.J. and Blakey, M.L., Seizing intellectual power at the New York African Burial Ground, *Historical Archaeology*, 31, 83–106, 1997.

Lester, D., Suicide and homicide among Native Americans: a comment, *Psychological Reports*, 77(1), 10, 1995.

Lewontin, R., The apportionment of human diversity, *Evolutionary Biology*, 6, 381–398, 1972.

Lichtenstein, B., Stigma as a barrier to treatment of sexually transmitted infection in the American Deep South: issues of race, gender and poverty, *Social Science and Medicine*, 57, 2435–2445, 2003.

Lippert, D., In front of the mirror: Native Americans and academic archaeology, in *Native Americans and Archaeologists: Stepping Stones to Common Ground*, Swidler, N., Dongoske, K.E., Anyon, R., et al., Eds., AltaMira Press, Walnut Creek, CA, 1997, pp. 120–127.

Lovejoy, C.O., Meindl, R.S., Mensforth, R.P., et al., Multifactorial determination of skeletal age at death: a method and blind tests of its accuracy, *American Journal of Physical Anthropology*, 68, 1–14, 1985.

Mack, M.E. and Blakey, M.L., The New York African Burial Ground Project: past biases, current dilemmas, and future research opportunities, *Historical Archaeology*, 38, 10–17, 2004.

Martin, D.L. and Frayer, D.W., *Troubled Times: Violence and Warfare in the Past*, Gordon and Breach, Amsterdam, 1997.

Martin, D.L. and Goodman, A.H., Health conditions before Columbus: paleopathology of Native North Americans, *Western Journal of Medicine*, 176, 65–68, 2002.

Mathe, B., Kaleidoscopic classifications: redefining information in a world cultural context, *International Journal of Special Libraries (INSPEL)*, 33, 56–60, 1999.

May, P.A., Alcohol policy considerations for Indian reservations and bordertown communities, *American Indian and Alaska Native Mental Health Research*, 4, 5–159, 1992.

McGinnis, S. and Davis, R.K., Domestic well water quality within tribal lands of eastern Nebraska, *Environmental Geology*, 41, 321–329, 2001.

Mellor, J.M. and Milyo, J.D., Individual health status and racial minority concentration in US states and counties, *American Journal of Public Health*, 94, 1043–1048, 2004.

Minior, T., Galea, S., Stuber. J., et al., Racial differences in discrimination experiences and responses among minority substance users, *Ethnicity and Disease*, 13, 521–527, 2003.

Peterson, J.E., Dance of the dead: a legal tango for control of Native American skeletal remains, *American Indian Law Review*, 15, 115–150, 1990/1991.

Powell, J. and Rose, J.C., Report on the osteological assessment of the 'Kennewick Man' skeleton, *Summary of Scientific Inquiries into Kennewick Man*, National Park Service Archaeology and Ethnography Program, U.S. Department of the Interior, 1999.

Price, T.D., Manzanilla, L., and Middleton, W.D., Immigration and the ancient city of Teotihuacan in Mexico: a study using strontium isotope ratios in human bone and teeth, *Journal of Archaeological Science*, 27, 903–913, 2000.

Ravesloot, J.C., Changing Native American perceptions of archaeology and archaeologists, in *Native Americans and Archaeologists: Stepping Stones to Common Ground*, Swidler, N., Dongoske, K.E., Anyon, R., et al., Eds., AltaMira Press, Walnut Creek, CA, 1997, pp. 172–178.

Roberts, D.G. and McCarthy, J.P., Descendant community partnering in the archaeological and bioanthropological investigations of African American skeletal populations: two interrelated case studies from Philadelphia, in *Bodies of Evidence: Reconstructing History through Skeletal Analysis*, Grauer, A.L., Ed., Wiley-Liss, New York, 1995, pp. 19–36.

Rose, J.C., Green, T.J., and Green, V.D., NAGPRA is forever: osteology and the repatriation of skeletons, *Annual Review of Anthropology*, 25, 81–103, 1996.

Rydell, R.W., *All the World's a Fair, Visions of Empire at American International Expositions, 1876-1916*, University of Chicago Press, Chicago, 1987.

Sankar, P., Cho, M.K., Condit, C.M., et al., Genetic research and health disparities, *JAMA*, 291(24), 2985–2989, 2004.

Santosham, M., Sack, R.B., Reid, R., et al., Diarrheal diseases in the White Mountain Apaches: epidemiologic studies, *Journal of Diarrheal Diseases Research*, 13, 18–28, 1995.

Sauer, N.J., Forensic anthropology and the concept of race: if races don't exist, why are forensic anthropologists so good at identifying them?, *Social Science and Medicine*, 34(2), 107–111, 1992.

Schoeninger, M.J., Hiebert, K., and Vandermerwe, N., Decrease in diet quality between the prehistoric and the contact period on St. Catherine's Island, Georgia, *American Journal of Physical Anthropology*, 72(2), 252, 1987.

Smay, D.B., Galloway, A., and Mason, R.T., A Possible Case of Gender Role Reversal, paper presented at the American Association of Physical Anthropologists National Meetings, Salt Lake City, UT, 1998.

Steckel, R.H. and Rose, J.C., *The Backbone of History: Health and Nutrition in the Western Hemisphere*, Cambridge University Press, Cambridge, 2002.

Steyn, M., A comparison between pre- and post-colonial health in the northern parts of South Africa: a preliminary study, *World Archaeology*, 35, 276–288, 2003.

Stocking, G.W., Franz Boas and the founding of the American Anthropological Association, *American Anthropologist*, 62, 1–17, 1960.

Thomas, D.H., *Skull Wars: Kennewick Man, Archaeology and the Battle for Native Identity*, Basic Books, New York, 2000.

Tsosie, R., Indigenous rights and archaeology, in *Native Americans and Archaeologists: Stepping Stones to Common Ground*, Swidler, N., Dongoske, K.E., Anyon, R., et al., Eds., AltaMira Press, Walnut Creek, CA, 1997, pp. 65–76.

Ubelaker, D.H., Katzenberg, M.A., and Doyon, L.G., Status and diet in precontact Highland Ecuador, *American Journal of Physical Anthropology*, 97, 403–411, 1995.

White, T.D. and Folkens, P.A., *Human Osteology*, Academic Press, San Diego, CA, 2000.

White Deer, G., Return of the sacred: Spirituality and the scientific imperative, in *Native Americans and Archaeologists: Stepping Stones to Common Ground*, Swidler, N., Dongoske, K.E., Anyon, R., et al., Eds., AltaMira Press, Walnut Creek, CA, 1997, pp. 37–43.

Wolf, E.R., *Europe and the People without History*, University of California Press, Berkeley, 1982.

Zimmerman, L.J., Remythologizing the relationship between Indians and archaeologists, in *Native Americans and Archaeologists: Stepping Stones to Common Ground*, Swidler, N., Dongoske, K.E., Anyon, R., et al., Eds., AltaMira Press, Walnut Creek, CA, 1997, pp. 44–56.

12

SEEING CULTURE IN BIOLOGY

Alan Goodman

CONTENTS

INTRODUCTION

The chapters in this volume attest to the complexity of study, interpretation, and sociopolitical significance of human biology. In complementary fashion they illustrate that what we take to be facts derived from biologicals — things such as genes, bones, and blood — are less stable than we might have realized. Complicating matters, biologicals are immersed in ever-changing biophysical, cultural, and sociopolitical contexts. Developmental and evolutionary histories of biologicals such as sickle cell anemia and diabetes or, as I present in this paper, skin color and height, are intertwined with sociopolitical histories and histories of study. When dealing with humans, the biologicals are also *culturals* in that their study, development, and evolution involve interrelated ideological, social, and political economic processes that were there from the very beginning.

All biological organisms develop in what Richard Lewontin (2000) has metaphorically referred to as the "triple helix" of genes, organisms, and environments. In studying humans, the chapters in this book show the need for taking seriously a fourth and particularly frayed and fuzzy helical strand: culture. Like water being the context in which a fish swims, culture is our context. Like water, too, culture is not just "out there" as context, milieu, and ecological niche. It is literally also in our biology. To be more precise, biologicals are actually *bioculturals*.

This chapter starts with a brief overview of the development of biological anthropology in North America. My purpose is to illustrate how biological anthropology is positioned between biological and social fields of study. I propose that the choosing of sides, specifically, placing the emphasis on biology to the strong relegation of anthropology, while an understandable coping strategy, is ultimately short sighted. It is giving up too soon on productive biocultural tensions. The chapters in this volume, written by social and biological scientists from both sides of the Atlantic, attest to the importance of combining cultural and biological analyses.

The biocultural tensions and the necessity of seeing culture in biology are further illustrated by two brief case studies: race and height. Rather than focus on the critique of the idea of race as a way to classify humans biologically, in the first case study I ask: "If race is so clearly not genetic, why do scientists and other groups largely continue to see race as biologically based?" The answer, I propose, requires an understanding of the cultures of science and the interrelationship of science and society.

The second case concerns how we interpret and come to understand the significance of another commonly observed phenotype: height. As a biosocial signifier and object of study, height (or stature) was once thought to be inherited, and then with the rise of nutritional anthropometry, height shifted meaning and became a powerful indicator of impoverished environments. Height first embodied nature and then nurture. Now, height remains an indicator of individual breeding, vitality, and desirability (whether owing to genes or environment). However, one can increase height either by limb lengthening or growth hormone injections, noteworthy because the change occurs without changing either the genes or environments that originally contributed to height. What does height embody in this context? In this chapter, I highlight the changing etiological context of height and the bioethics of changing individual heights within the normal range of variation by medical intervention.

THE DEVELOPMENT OF BIOLOGICAL ANTHROPOLOGY

Particularly in North America, the field of biological or physical anthropology has long been regarded as simultaneously a subfield of anthropology and a separate entity in its own right. In 1918, the *American Journal of Physical Anthropology* (*AJPA*) was first published under the editorship of Aleš Hrdlička of the Smithsonian Institute. In contrast to already existing journals such as the *American Anthropologist* (*AA*), published by the American Anthropological Association (AAA) with a mandate to cover all aspects of anthropology, *AJPA* focused exclusively on human biology, past and present. Soon after the initiation of the *AJPA*, E.A. Hooton, a classically trained biological anthropologist, took up his post at Harvard and began to turn out North America's first graduate students with a clear focus in biological anthropology. A decade later, in 1928, the American Association of Physical Anthropologists (AAPA) was formed and adopted the *AJPA* as its official journal. From this time on, U.S. biological anthropologists had two associations to affiliate with, AAPA and AAA, and two journals, *AJPA* and *AA*, to purchase, publish in, and read. The two associations with their respective meetings and journals represent the two foundational legs of biological anthropology, with one planted in anthropology and the other set upon biology.

Despite the separate journals and associations and the differences in objects of study, biological anthropology was becoming a part of anthropology by the middle of the twentieth century. Under the Boasian influence, many North American anthropologists were becoming trained as, and considered themselves to be, generalists or four-field anthropologists with expertise in more than one of the classic subfields of anthropology: biology, culture, linguistics, and archaeology.

The centripetal moment did not last very long, and some suggest that it is even part of a sort of disciplinary mythology (Segal and Yanagisako, 2005). With increased specialization and technological development in the biological sciences, and then with theoretical splits within anthropology in the late 1980s and 1990s, biological anthropology fit increasingly uneasily between anthropology (humanities and social sciences) and the biological sciences (biomedicine, genomics, and evolutionary theory). Tensions increasingly arose as the parent fields became more specialized and then drifted apart in theory and practice, questions and concerns. The result is that in the twenty-first century many biological anthropologists are emphasizing biology over anthropology: biomedicine, genetics, and related fields are perceived to be more stable, concrete, and noticeably better funded.*

* Biological anthropologists are also beginning to establish separate departments and often move out of anthropology departments. In May 2005 a proposal was put forward to split Harvard University's Department of Anthropology. This North American situation is beginning to look more like the organization of human biology in Europe, where separate departments of biological anthropology, archaeology, and cultural anthropology are common. What is different from at least the British tradition of human biology is that despite separate institutional homes, in my view many human biologists in Europe have retained a very strong sense of culture.

Based on membership figures, in 2005 it appears that far more physical anthropologists, perhaps as many as four to one, are members of the AAPA than are members of the Biological Anthropology Section (BAS) of the AAA. Uncomfortable with legs in two fields that are spread apart, many biological anthropologists have taken a few toes, if not the full leg, out of anthropology.

For comparison, in the 1980s and 1990s American archaeology was suffering from similar theoretical and institutional tensions about whether to remain in anthropology or go a separate way. The defense of archaeology-as-anthropology was strong (Nichols et al., 2003). Philip Phillips (1955:246–247), one of the first advocates of anthropological archaeology, succinctly stated that "New World archaeology is anthropology or it is nothing."*

Although the place of biological anthropology in anthropology has probably been even more tenuous than the place of archaeology in anthropology, the tensions, unfortunately, have not been nearly as well studied. Following Phillips by a half century, I propose that *biological anthropology is nothing if it is not anthropology*. Without grounding in the social sciences, biological anthropology is at risk of being merely a derivative field of biology. However, by maintaining an anthropological lens, biological anthropology can illustrate how to bridge bodies of knowledge and differing perspectives, leading to deeper understandings and a more ethical science. The following discussion provides two illustrations of the vitality of seeing culture in biology: in the study of biology, in development, and in evolution.

IF RACE IS NOT BIOLOGICALLY REAL, THEN WHY DO WE STILL ACT AS IF IT IS?

Race as Biologically Real

Parts of the interwoven social and academic histories of the idea of race have been told well by a number of authorities. For example, Tucker (1994) focuses on the science and politics of race in the field of psychology and intelligence testing, while Baker (1998) does much the same for anthropology, and others such as Barkan (1992) and Smedley (1999) have presented histories that are more general.

While there is healthy variation in opinion about how the idea of race came to be so thoroughly reified, there is a consensus on the basic outline of the story, which goes something like this. The idea of race was born out of the *scala naturae* and typological and ideological thinking. The precept of a natural and unchanging hierarchy of types of humans was

* Interestingly, the debate over the place of archaeology in academia has waxed and waned but has never subsided (see Gillespie and Nichols, 2003).

key to the worldview that allowed for and maintained the idea of race. Race, once theoretically and scientifically real, became a powerful ideological tool for the justification of colonialism and, not inconsequentially, of slavery.

Racial science reached full throttle in the United States in the mid-nineteenth century, ostensibly proving that races were separate entities, even separate species, and unequal in abilities (Smedley, 1999). With hindsight, it is not difficult to see that scientists such as George Morton, famed for his collection and measurement of skulls of ancient and modern races, were operating within a set of preconceived and politically expedient ideas (Gould, 1981).

However, the appeal of a politically useful idea is even more strongly illustrated by the contemporary continuation of scholarship that treats races as separate and differently talented (Herrnstein and Murray, 1994; Rushton, 1994; Sarich and Miele, 2004). The idea of race should have been blistered and torn by Darwinian attack on fixed types in the late nineteenth century. Then, data began to be compiled to show that race did not explain human variation; rather, variation was continuous and nonconcordant. By 1972, when Lewontin published his study of the apportionment of human variation, genetic data joined theory in effectively proving that race did not explain human variation (Lewontin, 1972; also see Goodman, 1997). Since 1972, an accelerating pace of genomic information from around the globe has smashed the idea that human genetic diversity can be usefully reduced to racial categories (Templeton, 2003; Yu et al., 2002). It should be patently obvious that in its applications in areas such as forensic anthropology and biomedicine, the reduction of human biological variation to race is more likely to lead to mistaken diagnoses, erroneous identifications, and inappropriate treatment plans than correct ones (Goodman, 1997; Stevens, 2003).*

Why then does the public still think race is biologically based, as do many scientists as well? When it comes to race, everyone seems to be dazed and confused (Goodman, 1997). Most scholars of scientific racism agree that race remains with us as a bio-scientific category because of the continued work that the concept does for the socially or politically powerful, because of the conservative nature of science, and because race is everywhere and obvious (Barkan, 1992; Fields, 1990). Tatum (1997) adds that we live in racial smog; that is, race and racism are all around us as part of our culture, ideas, and institutions, and like smog, it is hard to escape.

* Indeed, as this chapter was written, the U.S. Food and Drug Administration (FDA) approved BiDil® (Isosorbide Dinitrate and Hydralazine Hydrochloride) as a medicine for heart failure in one group: African Americans. It is the first such race-specific approval.

Tatum (1997) makes the point that in order to break out of a racialist culture, one needs to actively resist, but few make this decision and act on it, and the same is true for race in scientific culture. At least one lesson scientists can take from history is that what they say will first be shoehorned into existing ideologies and power structures. Stevens (2003) recently suggested that race is like a hammer. In the hands of scientists, it is as if every problem were a nail asking for a crude pounding instrument.*

Race as Biologically Unreal

Despite seeming to have both scientific facts and social justice on their side, scientists who have seen race as biologically unreal have not won the ideological war. I suggest that they have not for two reasons: first, because the scientists underestimated the power of racial ideology and the conservativeness of scientific culture and, second, because the scientists, social activists, and humanists have not come together to teach each other about the science of human variation and the cultural politics of race.** The following illustrates why I take this dim view.

Following in the long tradition of antiracist scholars including Frederick Douglas, Anterior Fermin, W.E.B. DuBois, Franz Boas, and Ashley Montagu, in the past decade an increasing number of scientists have begun to argue against race-as-biology for a general audience. In biological anthropology, C. Loring Brace, Leonard Lieberman, George Armelagos, Michael Blakey, and Jonathan Marks, among others, have led the intellectual battle, and they have been joined in this effort by the strong public voices of evolutionary biologists such as Richard Lewontin and Joseph Graves, the late Steven J. Gould, and genomic scientists such as Craig Venter, founder of Celera Genetics. All of these scientists (a) see race

* Many of the advocates of using race as a genetic grouping acknowledge that it is a crude tool, but the best we have available to us. For example, Steven E. Nissen, the chair of the FDA Advisory Committee for the approval of BiDil® opined "What we're doing here … is using self-identified race as a surrogate for genomic-based medicine." "I wish we had a gene chip."

** We recently asked a group of college students if race was based in part on biology. The majority answered "no." When I then asked the majority to explain why race was not biological, they replied that race was a "social construct." Pressed a step further, none of the "social constructionists" could explain the facts and observations of the group that proposed that race was based in part on biology. For example, they could not effectively counter an argument that high black infant mortality was due to genetic differences among so called races or that sickle cell anaemia was a "racial disease." As important as they are, humanistic intent, a politics of inclusion, and social understanding, are not sufficient to dismantle racial thinking. Science is also needed.

as a social reality and (b) embrace the study of human biological variation, but (c) argue against equating the tired and worn out idea of race with human biological variation.

As convincing as their arguments are, one is left to wonder how many individuals are reading their proclamations that race-as-genetics is dead in the science sections of newspapers and magazines, and for those who do read the articles, I wonder how this information is taken in against a lifetime of seeing the world through racial smog (Tatum, 1997). My guess is that few see the paradigm shift as logical, possible, or beneficial. It reminds me of a line from *Slaughterhouse Five,* by Kurt Vonnegut: "I was a student in the Department of Anthropology. At that time, they were teaching that there was absolutely no difference between anybody. They may be teaching that still."

Intentionally or not, in writing this line, Vonnegut shows that what scientists might think of as a simple matter, specifically that human biological variation exists but race does not explain it, is confusing to those who have long thought that race and human variation are the same thing. Two vignettes, one about a major news story and the other about a museum exhibit, further illustrate how just saying something different about race and variation is unlikely to be convincing.

On June 26, 2000, the idea that race has no biological basis became front-page news. The news event was the announcement of a working draft of the human DNA sequence. At the June 26 press conference celebrating this achievement, President Clinton, flanked by Craig Venter, the head of Celera, and Francis Collins, director of the National Institute of Health's (NIH) institute that oversees the Human Genome Project (HGP) in the United States, announced another finding. Celera and NIH had studied the DNA of individuals of different races and found them all to be remarkably similar. They proclaimed the news: genetically, there are no races. The pronouncement was momentous because of the political stature of the president and the flanking icons of science and industry. The press conference was featured on the front page of the *New York Times* and in many other well-circulated newspapers and popular magazines.

Yet, despite the sophisticated research, the well-placed publicity, and the power of presidential approval, racial profiling still goes on in medicine (see Satel, 2002; and Ashcroft, this volume) and race is still used as a biological category in fields such as forensic anthropology and medical genetics (see Outram and Ellison, this volume). Meanwhile, inhabitants of North America continue to believe what they think their eyes tell them: races are real and different.

Andre Langeny and his human biology colleagues in Switzerland had already realized that it would take more than a brief flourish of front-page news to break the hegemony of racial thinking. In the late 1990s they developed an impressive museum exhibit designed to teach the public about human variation and, by extension, the problematic nature of race-as-biology. The original exhibit remains at the Museum of Man in Paris, and an abridged version with 18 panels has been translated into English and brought to North America by Marshall Segall.* One panel illustrates how both tall and short peoples are found among Africans and Europeans; another panel illustrates the continuous distribution of skin colour; in another one can compare genetic similarities among different ethnic groups; yet another illustrates the nonconcordance of genetic traits.

The exhibit is wonderfully clear as to the nonracial nature of human biological variation, but nothing is said about racializing, race as an idea, or the connections between racializing and power. Race, it would seem, is merely a scientific concept. Scientists can argue scientifically for and against the reality and utility of race. It is all about the science.

To make a finer point about what is missing, the second panel of the exhibit combines two pictures. The smaller is a group of individuals dressed and the larger is a picture of the same individuals undressed. As the individuals in the second photo have left their clothing behind, the surrounding text exhorts the exhibit viewer to "leave culture in the cloakroom" in order to better see human phenotypic variation. This idea of *leaving culture in the cloakroom* is useful in the sense that it asks the viewer to try to see phenotypes without a racial lens. Such a shifting of vantage points is exactly what is needed to break through racial smog, and in a sense, this is what we need to do and what the exhibit helps the viewer to do. Biology and science are just part of the story, though. Where the exhibit stops is in putting the clothing back on. Armed with new understandings about race and biological variation, what does one do, and specifically how does one deal with the complicated, real biocultural world?

The question that needs to be asked is whether individuals who have seen the exhibit were changed by it. My sense is that if they fail to see how the idea of race-as-biology came about and lives on in their complicated biocultural world, the exhibit will remain merely academic. It will have been about science, but not about them. I suggest that the next steps for individuals and scientists require some understanding of history and social theory. To go a step further, we might not want to take those clothes off in the first place, because we need to better understand how clothing (i.e., culture) is intertwined with biology.

* See: http://allrelated.syr.edu/, accessed January 25, 2006.

HEIGHT AND HISTORY

Height (or stature) is both a core measure of human biology and a floating signifier, a trait that stands in for something else in an unstable way. The measurement of stature has long been among the most basic and frequently employed in the toolkit of human biology (Bogin, 1999; Tanner, 1981), but what are the possible meanings of height? In the following, I suggest that the meaning (or signification) of height has shifted in three intellectual and sociopolitical contexts. In the first context, height embodies status and breeding, in the second height embodies a better environment, and in the third, height is argued to embody better psychological well-being and improved earning.

Context 1: Height as Nature

When the science of anthropometry began in the eighteenth and early nineteenth century, height was one of the first measures to be included. The interpretations of measures such as greater height, weight, skull circumference, facial height, and skull volume were probably derived from previously observed connections. Anthropometric measurements of individuals and groups were initially thought to be inherited and unalterable, as exemplified by the use of anthropometrics in racial analyses (Gould, 1981). In the early twentieth century, average differences between individuals from different ethnic groups and countries appeared to be large. For example, the Maya, the Japanese, and Aboriginal Australians were small, and it was generally assumed that they were small because they were a (genetically) small (and some at this time would even say childlike) people (see also Santos, this volume).

Just as light skin tone became reified as a marker of race, and by extension better breeding, in Europe, tallness came to signify greater worth and wealth. Adults are taller than children, humans are taller than chimps and monkeys and most (but not all) animals, men are generally taller than women, Europeans of the age of exploration were typically taller (but not always) than non-Europeans, and upper-class Europeans such as land owners and the aristocracy were usually (but not always) taller than peasants and the landless. In all these cases, individuals with greater wealth are taller than those of lesser means. Going back to the eighteenth century, 80% of the taller of the U.S. presidential candidates won elections. With all of the aforementioned associations, it is not surprising that the height of individuals and groups became naturalized and embodied as a signifier of better breeding, race, and class (Bogin, 1999; Tanner, 1981).

Context 2 Height as Nurture

Then the field of international nutrition came along. In the early part of the twentieth century, height became re-inscribed with a different set of meanings. Early anthropometrists, including Franz Boas, began to realize that anthropometric traits, and especially traits that incorporated aspects of general body size (as opposed to skull shape metrics), were affected by food intake and environment. Rather than being a stable feature of different races and types, height was plastic. The effect of different environments was written into height. Individuals and groups were tall or short because of nurture rather than nature.

The inherent shortness and tallness of different groups was challenged mostly because of "natural experiments" such as Boas' (1912) observations that the heights of immigrant children born in the United States were greater than those born outside of the United States. Other observations were drawn from the adoption of many Aboriginal Australian children by white Australian families and the migration of many Japanese to the United States. In both cases, the height-for-age of the children from the assumed-to-be genetically short-statured group was closer to the reference standard for well-fed Western children (Habicht et al., 1974). Combined with clinical studies and other sources of information, a consensus emerged in the 1970s that height differences (at least before puberty) between *groups* growing up in different environments are best attributed to environmental differences rather than different genetic potentials for growth (Bogin, 1999; Steckel, 1995).*

The implications of this new environmental context–dependent worldview are tremendous. Seeing that group differences in size were due to environmental conditions allowed for the development of sensitive anthropometric measures of under-nutrition, such as changes in percentile height-for-age and weight-for-age, and the possibilities of earlier intervention to prevent individuals from becoming even more undernourished One could apply a universal standard to nearly every group in the world and come up with a metric of relative environmental deprivation or enrichment. Historical events and periods such as slavery could be "objectively studied" through anthropometric information (Komlos, 1995; Steckel, 1995). The success of development programs could be evaluated based on how well they improved children's growth. Finally, the slow growth of groups such as Cambodian refugees in the United States and Mayans in Mexico and Guatemala no longer could be attributed to genetics, but

* Conversely, in a shared environment and when environmental differences in food intake, disease load, work output, stress, and the like were minimal, differences in height among *individuals* may be best explained by genetic variation (Bogin, 1999).

signalled impoverished conditions. Shortness is still undesirable, but its origins are in the environment rather than in the genes.*

In this second phase, height became an embodied signifier of injustice. Smallness was something to fight against and correct. Doing so meant that a group was more likely to survive and prosper and to have a better quality of life. From an international aid worker's perspective, improving the growth of children in a targeted group signified that the group was on the road to improved health and wealth.

Context 3: The Medicalization of Shortness**

When it comes to stature, something new is in the works. To make a very bad pun, we have discovered some short cuts to an improved appearance of good health. Elliott (2003) aptly says we are looking to be "better than well." In 2004, the United States, the world's richest nation, spent 15% of its GNP on health care. There are many reasons why, but one of them is certainly the number of conditions that are medicalized, including short stature.

Middle- and upper-class individuals who are short in North America are not short because they do not get enough to eat, nor are they otherwise impoverished. Rather, they are short because of individual variation. Some

* In the 1980s, the "small but healthy" debate put the view that the height of groups of children could be used as a barometer of impoverishment to the test. Sekler set off the debate by proposing that stunted populations from the Indian subcontinent were small because they had positively adapted to their conditions of marginal food intake (Sekler, 1980; 1982). They were, in his adapted phrase, "small but healthy." Sekler argued that their smallness was not negative but an adaptation to poor nutritional conditions. The political-economic consequences of his proposal were clearly presented: if Indians are in fact *small but healthy*, then government and international aid ought to be directed to the fewer individuals with clinically clear signs of malnutrition versus the many millions of Indians who are small (Sekler, 1982). In response, international nutritionists and anthropologists showed that smallness, in the context within which it was then being studied, namely among peasants in the so-called developing world, was associated with, and in actual practice was an indicator of, functional costs such as lowered work capacity and compromised resistance to infectious disease (Allen, 1984). Because body growth is one of the first functions affected by nutritional inadequacies, it is a sort of miner's canary, an early indicator of the potential for serious biological problems. Reynaldo Martorell (1989) reviewed the available literature and summarized that smallness invariably led to increased mortality, so to argue that smallness was adaptive was equivalent to arguing that mortality is an adaptation. Rather than *small but healthy,* Martorell showed that a *bigger is better* view of height makes sense when environmental stressors are great and nutritional resources during development are limited.

** This section is inspired by and relies heavily upon Kimbrell (1997) and a senior honors thesis by Naomi Azar (2001).

individuals are short because of familial genetics and others may have gotten a mix of alleles that leads to small body size. The process of slow growth is hormonally mediated, and some individuals have identifiable deficiencies in the hormonal cascade that promotes growth, including a deficiency in the production of a wide variety of growth factors, including somatotropin. Shortness is not abnormal in these cases and it has no inherent biological relationship to health and well-being. Shortness is associated with neither better nature, as in the first context, nor better nutrition, as in the second context. However, even though the context has changed, the social significations have not disappeared. For these reasons shortness is seen as undesirable, and the perception has consequences for social status and biological well-being.

In the 1980s, somatotropin, a 2.2-kDa protein composed of 191 amino acids and popularly know as growth hormone (GH), became readily available after it was commercially synthesized and produced. Along with insulin, it was one of the first and greatest successes of bioengineering. With little testing as to its safety or efficacy as a height booster, synthetic human GH was approved by the U.S. Food and Drug Administration (FDA) for individuals with GH deficiency.* Synthetic human GH (HGH), marketed by major pharmaceutical companies under names such as Nutropin and Protropin (Genentech), and Humatrope (Eli Lilly) began to be prescribed as a viable treatment for the stigmatized "disease" of short stature.

The companies that produced versions of synthetic HGH initially had a limited market for their product because a relatively small number of individuals had been diagnosed as having HGH deficiency. The market needed to be made for synthetic HGH. Barriers to building a market among individuals with diagnosed GH deficiency and especially to the many others with idiopathic short stature were considerable. A prominent one was the difficult course of treatment, including daily injections for from 2 to 11 years at an average cost of over $15,000 per year (Kimbrell, 1997). In addition, there was little proof that HGH injections, injections of a very powerful anabolic hormone with a wide range of activities, were free of any long-term health risks.** Finally, there is still little evidence of efficacy as a growth promoter. The best available data, from studies supported by pharmaceutical companies, based on small sample sizes,

* Because of its pulsate and circadian rhythm, GH deficiency is difficult to diagnose. This fact has lead to considerable imprecision in identifying the degree of deficiency and individuals who actually suffer from GH deficiency. Similarly, because GH secretion represents the end of a hormonal axis and, as is typical of hormone systems, is involved in many feedbacks, its effects are difficult to isolate.

** Growth hormone is sold on a wide variety of websites. Here, on the virtual black market, it is purported to do many different things including build muscle mass, increase virility and prevent aging. It is mostly marketed to adults.

and subject to strong positive reporting biases, suggest that HGH provides an extra 5 to six 6 cm to final heights of both GH-deficient children and children with idiopathic short stature (with a chance of no improvement at all; Finkelstein et al., 2002). The findings of similar levels of growth enhancement in both GH- and non-GH-deficient children suggest that something other than the GH may be causal.* Finkelstein et al. (2002) estimate that current treatment costs are about $35,000 per inch.

Despite the considerable barriers, pharmaceutical companies have sold a lot of synthetic HGH. Some evidence has come to light of companies that have pushed families into taking synthetic HGH. Genentech reportedly paid 10 thousand dollars a year to the wife of a pediatric endocrinologist in North Carolina to find short children and refer them to her husband for evaluation (Kolata, 1994:1). A federal grand jury indicted a doctor David R. Brown of Minneapolis on charges that he received 1.1 million dollars in kickbacks from Genentech, Inc. for prescribing Protropin (Kolata, 1994:1). Back in the early 1990s, Brown was prescribing Protropin to 350 patients.

Wrongdoings aside, the main impetus for taking HGH appears to be a strong desire to be tall, or at least not short, and the desire is not entirely irrational. Although added height does not have any of the associated benefits it would have if height were restricted because of a poor diet, such as increased work capacity, height in the United States is associated with greater earning power and social prestige. Who would not want those?

In a sense, a cure of dubious efficacy has manufactured a disease. In a paradoxical flip of biosocial causality, the biological phenotype of height might still equate itself with wealth if only those with wealth can afford HGH. Herein lies a conflict between the individual and the group. Even if some individuals increase their stature through HGH, individuals with short stature will still be around. A new bottom will be established; only the distribution will be shifted up, and with the shift in distribution comes a shift in who is stigmatized.

The use of synthetic HGH in children with idiopathic short stature illustrates a conflict between an individual and a social logic. In July 2003, the FDA approved the use of Eli Lilly's synthetic HGH, Humatrope, for the treatment of individuals with idiopathic short stature. Who is to pay for the treatment? Should insurance companies pay, meaning that insurance premiums will rise for everyone who pays for insurance or should families pay? If the costs are born directly from the pay packets of families, then this enhancement will only be available to those who can

* For example, one would expect that the greatest improvement would be found in individuals with the lowest production of GH, but that does not appear to be the case.

afford it, thus turning again on its head the causal connection between height and wealth.

Of course, the old association of tallness with success is not limited to North America. Limb lengthening procedures are popular in China (Smith, 2002). Today, the Chinese are placing lower limits on the height of diplomats, 5 feet, 7 inches for males and 5 feet, 3 inches for females and, apparently, Chinese men have developed a strong preference for tall women. As a result, limb lengthening is increasing in popularity. In the "Llizarov procedure" (named after a Russian doctor), both legs are severed and a device is inserted that slowly separates the lower leg bones. As it does so, new bone is laid down.* A successful operation may add a few inches but could leave the patient with a permanent limp.

Although the discussion of administering growth hormone injections and limb lengthening might seem to be removed from what human biologists do, I think we have a role to play. One of the most important intellectual advances we can make is to point out places where culture and biology may be clashing. Steven Boyden (1970) would call limb lengthening and the administration of HGH injections "biocultural maladaptations." They are changing a phenotype without changing underlying conditions. It is changing an individual, at a cost to everyone, but not bettering life for all.

Regardless of how one achieves height, height is biological and cultural. Height is cultural in that in all phases height is an embodiment of socioeconomic contexts. Height is cultural in that in all phases height signifies something about social status. Finally, height further affects life chances and opportunities.

CONCLUSIONS

The purpose of this chapter has been to examine what is lost if we do not make biological anthropology fully anthropological, by which I mean taking culture seriously. Culture, I argue, is not a box to put things in, nor is it a one-dimensional variable, nor is it merely a context for human biology. In both the ways we think about nature and biology and in what we see and measure as natural and biological, culture is also *in* nature and biology. In an era of ever-increasing specialization, biological anthropology and human biology can offer something uniquely valuable: it is a natural born interdisciplinary field, one that knits together the social and the biological in a deeper analysis of human development, adaptation, variation, and behavior.

* This procedure was briefly illustrated in the film GATACA. The main character undergoes limb lengthening so as to pass for another.

Thirty years after an oft-cited aphorism, "archaeology is nothing if not anthropology," I propose that biological anthropology is nothing if it is not anthropological. Without grounding in the social sciences, humanities, and philosophy, biological anthropology is at risk of being a second rate stepchild of biology. However, by taking culture seriously, biological anthropology can illustrate how to bridge different sources of knowledge and different perspectives.

I propose that the biological and cultural wings of anthropology can make a great vehicle of understanding through synthesis if we show how biology and culture are interwoven with genes and environments. I have suggested that the new cultural–biological anthropology should incorporate a commitment to: (a) better understand the ethics and political-economy of difference; (b) better understand the shifting cultural lenses through which we view differences and similarities; and (c) better understand how the individuals and groups we work with perceive and understand us and themselves.

If we agree that human biology is cultural, then we must understand both cultural and biological dimensions. This may require us to rethink our education and the structure of academia. Human biologists need to be familiar with social analysis, history, and philosophy, for example. It also means that generalizations will not be as easy to come by, but this is not an unfortunate consequence. It is, rather, a recognition of something that has always been true: human biology is culturally and historically contingent. The epidemiological results from one location and time are different from the epidemiological results from another location and time. This is not because of error or noise, but because the particular ways that illness develops and takes on meaning differ depending on the historical and cultural milieu. Class differences in health also vary from place to place because class is culturally and historically contingent. As I hope I have shown, the meanings and practices of race change over time and place. Not taking into account the cultural and historical context of racial ideologies and practices may doom to failure even the most proactive, creative, and best-intended scientific antiracialists. Today, height means something different to a girl growing up in the capital of China, a boy growing up in rural Tennessee, and a child growing up in poverty in southern India or the horn of Africa. The task of the human biologist is to recognize the biocultural similarities and the differences.

ACKNOWLEDGMENTS

Ideas expressed in this paper have been sparked and refined by conversations with friends and colleagues including George Armelagos (Emory University) and Chaia Heller and Lynn Morgan (Mt. Holyoke College).

I owe a large dose of gratitude to Naomi Azar. As a student at Hampshire College, Naomi completed a stellar honors thesis on growth hormone and the manufacturing of a disease, making my job that much easier. Joan Barrett helped edit the manuscript. Finally, I thank George Ellison for giving me the freedom and opportunity to try to say what I wanted to say.

REFERENCES

Allen, L., Functional indicators of nutritional status of the whole individual or the community, *Clinical Nutrition* 3, 169–175, 1984.

Azar, N., Disease, Treatment and the Market Place: Growth Hormone Therapy and the Manufacturing of a Disease, B.A. Thesis, Hampshire College, Amherst, MA, 2001.

Baker, L., *From Savage to Negro*, University of California Press, Berkeley, 1998.

Barkan, E., *The Retreat of Scientific Racism*, Cambridge University Press, New York, 1992.

Boas, F., *Changes in Body Form of Descendants of Immigrants*, Columbia, New York, 1912.

Bogin, B., *Patterns of Human Growth*, 2nd ed., Cambridge University Press, Cambridge, 1999.

Boyden, S.V., Cultural adaptation to biological maladjustment, in *The Impact of Civilization on the Biology of Man*, Boyden, S.V., Ed., University of Toronto Press, Toronto, 1970, pp. 190–218.

Elliott, C., *Better than Well*, WW Norton, New York, 2003.

Fields, B.J., Slavery, race and ideology in the United States of America, *New Left Review*, 181, 95–118, 1990.

Finkelstein, B.S., Imperiale, T.F., Speroff, T., et al., Effect of growth hormone therapy on height in children with idiopathic short stature, *Archives of Pediatric and Adolescent Medicine*, 156, 230–240, 2002.

Gillespie, S.D. and Nichols, D.L., Archaeology is anthropology, *Archaeological Papers of the American Anthropological Association*, 13, 155–169, 2003.

Goodman, A., Bred in the bone?, *Sciences*, March/April, 20–25, 1997.

Gould, S.J., *The Mismeasure of Man*, WW Norton, New York, 1981.

Graves, J., *The Emperor's New Clothes*, Rutgers University Press, London, 2001.

Habicht, J.-P., Martorell, R., Yarbrough, C., et al., Height and weight standards for preschool children: how relevant are ethnic differences in growth potential?, *Lancet*, April 6, 611–614, 1974.

Herrnstein, R. and Murray, C., *The Bell Curve*, Free Press, New York, 1994.

Hintz, R.L., Attie, K.M., Baptista, J., et al., Effect of growth hormone treatment on adult height of children with idiopathic short stature, *New England Journal of Medicine*, 340, 502–507, 1999.

Kimbrell, W., *The Human Body Shop*, Regnery, Washington, DC, 1997.

Kolata, G., Selling growth drug for children: legal and ethical questions, *New York Times*, August 15, 1994, pp. 1.

Komlos, J., Ed., *The Biological Standard of Living in Europe and America, 1700-1900: Studies in Anthropometric History*, Variorum, Aldershot, U.K., 1995.

Lewontin, R.C., The apportionment of human diversity, *Evolutionary Biology*, 6, 381–398, 1972.

Lewontin, R., *The Triple Helix*, Harvard University Press, London, 2000.

Martorell, R., Body size, adaptation and function, *Human Organization*, 48, 15–20, 1989.

Nichols, D.L., Joyve, R., and Gillespie, S.D., Archaeology must have anthropology, *Archaeological Papers of the American Anthropological Association*, 13, 3–13, 2003.

Phillips, P., American archaeology and general anthropological theory, *Southwestern Journal of Anthropology*, 11, 246–250, 1955.

Rushton, J.P., *Race, Evolution and Behavior*, Transactions, New Brunswick, NJ 1994.

Satel, S., I am a racially profiling doctor, *New York Times*, May 5, 2002.

Sarich, V. and Miele, F., *Race: The Reality of Human Differences*, Westview, Boulder, CO, 2004.

Segal, D. and Yanagisako, S., Introduction, in *Unwrapping the Sacred Bundle: Reflections on the Disciplining of Anthropology*, Segal, D. and Yanagisako, S., Eds., Duke University Press, Durham, NC, 2005, pp. 1–23.

Sekler, D., Malnutrition: an intellectual odyssey, *Western Journal of Agricultural Economics*, 5, 219–227, 1980.

Sekler, D., 'Small but healthy': a basic hypothesis in the theory, measurement and policy of malnutrition, in *Newer Concepts in Nutrition and Their Implications for Policy*, Sukharme, P.V., Ed., Maharashtra Association for the Cultivation of Science Research Institute, Pune, India, 1982, pp. 127–137.

Smedley, A., *Race in North America*, 2nd ed., Westview Press, Boulder, CO, 1999.

Smith, C., Risking limbs for height, and success, in China, *New York Times*, May 5, 2002, pp. 3.

Steckel, R.H., Stature and the standard of living, *Journal of Economic History*, 33, 1903–1940, 1995.

Stevens, J., Racial meaning and scientific methods: changing policies for NIH-sponsored publications reporting human variation, *Journal of Health Policy, Politics and Law*, 28, 1033–1087, 2003.

Tanner, J.M., *A History of the Study of Human Growth*, Cambridge University Press, New York, 1981.

Tatum, B., *"Why Are All the Black Kids Sitting Together in the Cafeteria?"*, Basic Books, New York, 1997.

Templeton, A., Human races in the context of recent human evolution: a molecular genetic perspective, in *Genetic Nature/Culture*, Goodman, A., Heath, D., and Lindee, S., Eds., University of California Press, Berkeley, 2003, pp 234–257.

Tucker, W.H., *The Science and Politics of Racial Research*, University of Illinois Press, Chicago, 1994.

Yu, N., Chen, F-C., Ota, S., et al., Larger genetic differences within Africans than between Africans and Eurasians, *Genetics*, 161, 269–274, 2002.

INDEX

C

D

H

N

O

P

R

S

T

For Product Safety Concerns and Information please contact our EU representative GPSR@taylorandfrancis.com Taylor & Francis Verlag GmbH, Kaufingerstraße 24, 80331 München, Germany

Batch number: 10725802

Printed by Printforce, the Netherlands